Piezoelectric Materials

WILEY SERIES IN MATERIALS FOR ELECTRONIC & OPTOELECTRONIC APPLICATIONS

www.wiley.com/go/meoa

Series Editors

Professor Richard Curry, University of Manchester, Manchester, UK
Professor Harry Ruda, University of Toronto, Toronto, Canada
Professor Jun Luo, Chinese Academy of Sciences, Beijing, China

Honorary Series Editors

Professor Arthur Willoughby, University of Southampton, Southampton, UK
Dr Peter Capper, Ex-Leonardo MW Ltd, Southampton, UK
Professor Safa Kasap, University of Saskatchewan, Saskatoon, Canada

Published Titles

Bulk Crystal Growth of Electronic, Optical and Optoelectronic Materials, *Edited by P. Capper*
Properties of Group-IV, III—V and II—VI Semiconductors, *S. Adachi*
Charge Transport in Disordered Solids with Applications in Electronics, *Edited by S. Baranovski*
Optical Properties of Condensed Matter and Applications, *Edited by J. Singh*
Thin Film Solar Cells: Fabrication, Characterization, and Applications, *Edited by J. Poortmans and V. Arkhipov*
Dielectric Films for Advanced Microelectronics, *Edited by M. R. Baklanov, M. Green, and K. Maex*
Liquid Phase Epitaxy of Electronic, Optical and Optoelectronic Materials, *Edited by P. Capper and M. Mauk*
Molecular Electronics: From Principles to Practice, *M. Petty*
Luminescent Materials and Applications, *A. Kitai*
CVD Diamond for Electronic Devices and Sensors, *Edited by R. S. Sussmann*
Properties of Semiconductor Alloys: Group-IV, III—V and II—VI Semiconductors, *S. Adachi*
Mercury Cadmium Telluride, *Edited by P. Capper and J. Garland*
Zinc Oxide Materials for Electronic and Optoelectronic Device Applications, *Edited by C. Litton, D. C. Reynolds, and T. C. Collins*
Lead-Free Solders: Materials Reliability for Electronics, *Edited by K. N. Subramunian*
Silicon Photonics: Fundamentals and Devices, *M. Jamal Deen and P. K. Basu*
Nanostructured and Subwavelength Waveguides: Fundamentals and Applications, *M. Skorobogatiy*
Photovoltaic Materials: From Crystalline Silicon to Third-Generation Approaches, *Edited by G. Conibeer and A. Willoughby*
Glancing Angle Deposition of Thin Films: Engineering the Nanoscale, *M.M. Hawkeye, M.T. Taschuk and M.J. Brett*
Physical Properties of High-Temperature Superconductors, *R. Wesche*
Spintronics for Next Generation Innovative Devices, *Edited by K. Sato and E. Saitoh*
Inorganic Glasses for Photonics: Fundamentals, Engineering and Applications, *A. Jha*
Amorphous Semiconductors: Structural, Optical and Electronic Properties, *K. Morigaki, S. Kugler and K. Shimakawa*
Microwave Materials and Applications, Two-volume set, *Edited by M. T. Sebastian, R. Ubic and H. Jantunen*
Molecular Beam Epitaxy: Materials and Applications for Electronics and Optoelectronics, *Edited by H. Asahi and Y. Korikoshi*
Metalorganic Vapor Phase Epitaxy (MOVPE): Growth, Materials Properties, and Applications, *Edited by S. Irvine and P. Capper*
Optical Properties of Materials and Their Applications, Second Edition, *Edited by J. Singh*
Oxide Electronics, *Edited by A. Ray*
Organic Semiconductors for Optoelectronics, *Edited by H. Naito*
Phosphors for Radiation Detection, *Edited by T. Yanagida and M. Koshimizu*
Spintronics: Materials, Devices and Applications, *Edited by Kaiyou Wang, Meiyin Yang and Jun Luo.*
Photoconductivity and Photoconductive Materials: Fundamentals, Techniques and Applications, *Edited by Safa O. Kasap*
Electromagnetic Wave Absorbing Materials: Fundamentals and Applications, *First Edition. Edited by Hongjing Wu, Jun Luo, and Meiyin Yang.*
Piezoelectric Materials: Properties, Applications and Devices, *First Edition. Christopher R. Bowen and Andrew R. Plummer.*

Piezoelectric Materials

Properties, Applications and Devices

Christopher R. Bowen
Department of Mechanical Engineering
University of Bath, UK

Andrew R. Plummer
Department of Mechanical Engineering
University of Bath, UK

WILEY

This edition first published 2026

Registered Offices
John Wiley & Sons, Inc., 111 River Street, Hoboken, NJ 07030, USA
John Wiley & Sons Ltd, New Era House, 8 Oldlands Way, Bognor Regis, West Sussex, PO22 9NQ, UK

For details of our global editorial offices, customer services, and more information about Wiley products visit us at www.wiley.com.

The manufacturer's authorized representative according to the EU General Product Safety Regulation is Wiley-VCH GmbH, Boschstr. 12, 69469 Weinheim, Germany, e-mail: Product_Safety@wiley.com.

Wiley also publishes its books in a variety of electronic formats and by print-on-demand. Some content that appears in standard print versions of this book may not be available in other formats.

Library of Congress Cataloging-in-Publication Data Applied for:
ePDF: 9781118572863
epub: 9781118572856
Hardback ISBN: 9781118572887

Cover Design and Images: Wiley

Set in 9.5/12.5pt STIXTwoText by Straive, Pondicherry, India
Printed and bound by CPI Group (UK) Ltd, Croydon, CR0 4YY

C9781118572887_301025

Contents

Preface

Piezoelectric materials are the most widely used type of smart material. They can be used to create both actuators and sensors, and they are increasingly important for energy harvesting. The purpose of this book is to make piezoelectric materials technology more accessible by crossing the boundary between fundamental materials physics and applied engineering. It will be of use to researchers, advanced students, and practicing engineers.

Our goal is to provide a comprehensive overview of the subject, covering material types, properties and devices, and including application examples in the form of in-depth case studies. Ferroelectrics are introduced as the type of piezoelectric material that has the widest practical usage in engineering, and the pyroelectric property is explained. The underlying science in Chapters 2 and 3 will help the reader understand the limits and appropriate operating conditions for piezoelectric materials. Chapter 4 describes piezoelectric actuators, including stacks, bending actuators and amplification mechanisms. Sizing equations and optimized design in the light of load behaviour is covered. The utilization of the piezoelectric effect to form sensors is explained in Chapter 5. Accelerometers, force and pressure sensors, and hydrophones are discussed in detail. Energy harvesting is attracting increasing interest since it provides a route to produce autonomous self-powered electronic devices. The emerging use of piezoelectric materials in energy harvesters is described in Chapter 6. Electronic circuits used for interfacing to piezoelectric actuators and sensors are introduced in Chapter 7, along with position sensing and controller design for closed-loop control of piezoelectric actuators. Chapter 8 describes several piezoelectric engineering applications and studies in detail, including use in valves, pumps, vibration control, porous materials for sensing, bi-stable structures, and self-heating effects.

Key points are highlighted throughout the text in the form of "notes". We also include some worked examples and lists of references for further reading at the end of each Chapter. To summarize, we hope this book will provide an appreciation of the fundamentals of piezoelectric materials, together with an understanding of how to use the materials to create practical engineering devices.

September 2025

Christopher R. Bowen
Andrew R. Plummer
Bath, UK

Acknowledgements

We thank all colleagues with whom we have interacted on this subject matter over many years, and who have thus influenced this book to a varying degree. In particular we would like to thank these current or former colleagues, students and collaborators: Dhinesh Sangiah, Johan Persson, Paul Guerrier, Ian Brooks, Phil Elliott, Hubertus Murrenhoff, JP Henderson, Mike Lyons, Michael Schlotter, Johan Berote, Warren Manning, Martin Levesley, Peter Giddings, Andrew Dent, Andrew Perry, Rhodri Lewis, Fran Baxter, Luke Nelson, James Roscow, Simon Panteny, Vana Adamki, Marcin Jan Kraśny, Daniel Zabek, Peter Harris, Jon Gittings, Mike Bertin, Yan Zhang.

1

Introduction

This book outlines the properties and applications of piezoelectric materials for sensor, actuator and energy harvesting applications. Piezoelectric materials are of significant interest since they exhibit both active and sensory responses, and can form the basis of *smart* or *multifunctional* devices, systems or structures. We will begin here with the basic definition of a smart system.

1.1 Active and Sensory Materials for Smart Systems

Before examining the properties and applications of piezoelectric materials, it is first of interest to define the terminology used to describe materials that are used as *actuators*, *sensors* or *energy harvesters* and how they are combined to create a *smart system*. We will describe an *active material* as a material that exhibits a specific response (which can be considered as an 'output') when some form of stimulus is applied (which can be considered as an 'input'). Specific examples of the potential stimuli/inputs include the application of force, an induced strain, vibrations, a change in temperature or the application of an electric or magnetic field [1]. Examples of the resulting responses/outputs to any stimulus by an *active material* can include a dimensional or shape change of the material, a change in its stiffness, the generation of electrical charge or even a change in colour. 'Classical' active materials, such as piezoelectric, shape memory alloys, dielectric elastomers or magnetostrictive materials [2–4], are widely employed as actuators in a smart system since they are able to produce a deflection, strain or force in response to an applied field (electrical, magnetic or thermal).

In addition to their use as an active element in an actuator, piezoelectric materials are attractive since they can also be employed as sensory elements, whereby any applied force or strain is converted into an electric charge, current or voltage. The generated signal can be used to measure the applied force or strain and provide a sensory function to a smart system. We will see later in this book that many piezoelectric materials are also pyroelectric, where a change in temperature also produces an electric charge, so these materials can often be employed as thermal sensors or even provide a combination of functions simultaneously. Piezoelectric materials are therefore particularly attractive since they can act as both an

Piezoelectric Materials: Properties, Applications and Devices, First Edition.
Christopher R. Bowen and Andrew R. Plummer.

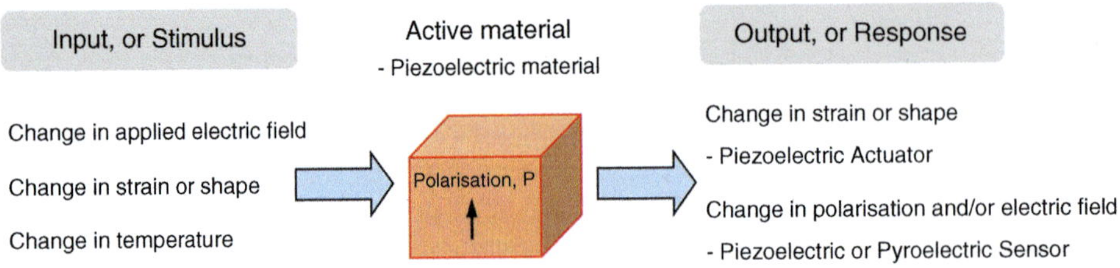

Figure 1.1 Schematic of a piezoelectric element employed as an *active material*, behaving as either a sensor or an actuator; the function depends on the nature of the stimulus and the resulting response. *Source:* Adapted from Nelson [1].

actuator and a sensor of mechanical and thermal stimuli, thereby providing a high degree of *multifunctionality*.

The concept of an active material that reacts to an applied stimulus (or an 'input') giving a resulting response (or an 'output') is outlined in Figure 1.1 for piezoelectric and pyroelectric elements.

In order to be able to react to any changes in its surroundings, a truly *smart system* should provide the following functions [1]:

i) the smart system must be able to receive sensory information from the active material in order to provide some form of 'sensing' function,
ii) the smart system must be able to process the sensory data and then provide an 'analysis and decision' function, and
iii) in response to the outcome of the 'analysis and decision' function, it must be able to act on that decision; for example, in a vibration control system, this can include the actuation of an engineering structure to reduce the level of vibration.

This hierarchy of a potential smart system is outlined in Figure 1.2 [1], which shows engineering structures exhibiting only an *actuated* or *sensory* function, along with *controlled*,

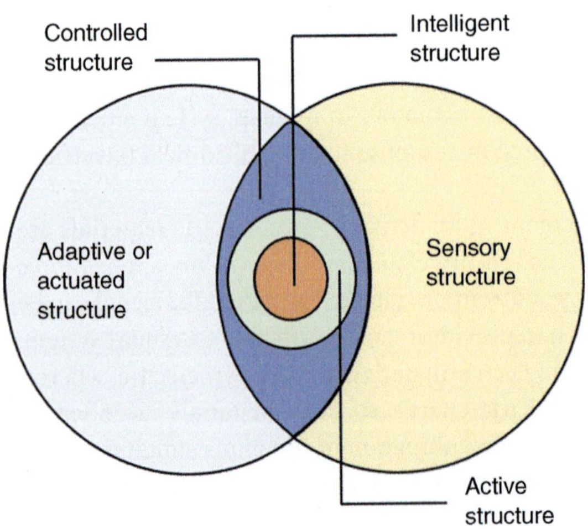

Figure 1.2 Representation of actuated, sensory, controlled, active and intelligent structures. A smart structure is typically a controlled, active or intelligent structure. *Source:* Adapted from [1–3].

active and intelligent *structures* that provide additional functionalities to the system. Potential applications for such smart systems include civil and transport structures [5–7], robotics [8] and wearables [9]. Based on Figure 1.2, engineering structures, which contain *only* sensors can be described as *sensory*. Structures that contain *only* actuators can be defined as *adaptive* or *actuated* structures. Engineering structures that have both sensing and actuation functionalities can be described as *controlled* structures if the ability to sense and then actuate is linked by some form of closed-loop control, for example, controlled passive vibration damping. Additional definitions of smart structures, such as *active* and *intelligent* structures, are indicative of a higher level of functionality; this can encompass a greater degree of integration, utilising external energy input (such as active vibration damping), the inclusion of additional power electronics or even exploiting energy harvesting [1–3].

1.2 Energy Harvesting Materials

In addition to sensing and actuation, there has been growing interest in the topic of energy harvesting in an effort to produce autonomous self-powered systems, for instance, self-powered structural health monitoring. This interest has been motivated by the reduction in the necessary power levels for wireless sensing and the continued growth in the Internet of Things (IoT), where large numbers of distributed sensors are to be deployed. A potential application for energy harvesting is to provide power for wireless sensor networks that consequently do not require any external power sources or any battery replacement; such networks often need only small levels of power in the μW–mW range [10–12]. The concept of 'energy harvesting' is therefore typically associated with scavenging local energy sources (vibrations, heat, airflow, etc.) to produce low levels of power and is therefore different from 'energy generation', which is associated with harnessing our primary sources of power on a large scale. Energy harvesting is of particular interest in the deployment of sensing elements or sensor networks in highly remote or hostile locations, or if many sensors are required [10]. The use of piezoelectric materials to transform vibrations, strains or dynamic forces into usable electrical energy, rather than simply acting as a sensing signal, is therefore a promising approach [10].

1.3 Multifunctional Materials, Devices, Systems and Structures

The research and development area of smart systems and structures is highly interdisciplinary, encompassing advanced materials, composites, electronics, mechanics, dynamics, physics and chemistry.

It is of interest to define here the terms *material*, *device*, *system* and *structure* since they are often used interchangeably in the literature. The active *material* can be considered as the simplest element, which can be a bulk material that simply reacts to a stimulus, as shown in Figure 1.1. At a higher level of complexity, *devices* that provide a sensing, actuating or harvesting functionality are then constructed from an active material. Complete engineering *systems* or *structures* are then designed that contain a combination of active, sensing and harvesting devices, along with some form of 'analysis and decision' function, usually in the form of a control system.

The concept of *multifunctionality* is also used, where the material or system can perform multiple tasks through the appropriate combination of different functional capabilities. This can be achieved using piezoelectric materials as they provide sensing, actuation, harvesting and even energy storage capability since they can also be used as simple capacitors [13, 14].

From our discussions above, it can be considered that it is difficult for a single material on its own to be truly smart, controlled or intelligent; it usually acts as an *active* material. As a result, the concept of a '*smart material*' is often used to describe a material that exhibits some form of *activity,* as shown in Figure 1.1, that allows it to be employed as an active element (sensor, actuator or harvester) in a smart system or structure.

Note 1.1

Smart systems or structures often combine a sensing capability with an analysis and decision function that leads to an active response, such as actuation.

Note 1.2

Multifunctional materials or systems have the ability to combine a range of functional capabilities such as sensing (force or temperature), actuation or energy harvesting.

Note 1.3

Smart systems have the potential to be self-powered by means of energy harvesting.

1.4 Piezoelectric, Pyroelectric and Ferroelectric Materials

We have seen that a piezoelectric material generates an electrical charge when subjected to a mechanical stress/strain, which can produce an electric field under open-circuit conditions or a current under short-circuit conditions. This type of response is termed the *direct* piezoelectric effect and forms the basis of the use of piezoelectric materials as an active material in a sensor device or an energy harvesting device. In addition, piezoelectric materials produce a strain (or displacement) when they are subjected to an applied electric field; this is termed the *converse* piezoelectric effect, which enables them to be used as an active material in an actuator device.

We will see in this book that *ferroelectric* materials represent an important subgroup of piezoelectric materials. The term *ferroelectric* is used to describe materials that exhibit a spontaneous polarisation, where the direction of polarisation can be reoriented (or 'switched') by the application of an externally applied electric field, and on the removal of that applied electric field there is a remnant polarisation that results in the material exhibiting piezoelectric behaviour. Ferroelectrics are piezoelectric materials that have the widest practical usage in engineering since they can be formed in a variety of geometries, and the polarisation direction can be selected and tailored by the application of an electric field after manufacture. We will see later that all ferroelectric materials are also *pyroelectric*, which means they can act as thermal sensors since they develop electrical charge when subjected to a change in temperature [15]. As an example, lead zirconate titanate (PZT) is a commonly used piezoelectric material that is also both ferroelectric and pyroelectric. This wide-ranging functionality makes these materials of particular interest for the range of structures in Figure 1.2.

This book is devoted to piezoelectric materials and their use in sensors, actuators and energy harvesters, either as discrete devices or as part of controlled, active or intelligent systems or structures with multiple functions. The properties of the materials will be discussed together with detailed examples of use in a wide variety of applications. The book attempts to cross the boundary between fundamental materials physics and applied engineering to ultimately provide engineers with a key understanding to help create either discrete sensors, actuators or energy harvesters or lead to the formation of more advanced smart systems. The book deliberately provides some of the underlying science to help those interested in using piezoelectric materials in engineering applications. It is important to examine these concepts since the underlying science in Chapters 2 and 3 will help the reader understand the limits and appropriate operating conditions for piezoelectric, ferroelectric and pyroelectric materials for use in the sensing, actuator and harvesting applications that are discussed in Chapters 4–8.

References

1 Nelson, L.J. 2005. Characterisation and modelling of active fibre composites. PhD Thesis. University of Bath.

2 Hodgkinson, J. (1993). What are smart materials anyway? *Materials World* 1: 449–451.

3 Crawley, E., Bent, A., Warkentin, D., and Hagood, N. (1994). Recent progress in intelligent material systems. *Materiaux and Techniques* 12: 31–37.

4 McEvoy, M.A., and N. Correll, N. (2015). Materials that couple sensing, actuation, computation, and communication. *Science* 347: 1261689.

5 Sun, J., Guan, Q., Liu, Y., and Leng, J. (2016). Morphing aircraft based on smart materials and structures: a state-of-the-art review. *Journal of Intelligent Material Systems and Structures* 27 (17): 2289–2312.

6 Chopra, I. (2002). Review of state of art of smart structures and integrated systems. *AIAA Journal* 40 (11): 2145–2187.

7 Han, B., Wang, Y., Dong, S. et al. (2015). Smart concretes and structures: A review. *Journal of Intelligent Material Systems and Structures* 26 (11): 1303–1345.

8 Hamed, Y., O'Donnell, G., Lishchenko, N., and Munina, I. (2023). Strain sensing technology to enable next-generation industry and smart machines for the factories of the future: a review. *IEEE Sensors Journal* 23 (21): 25618–25649.

9 Xie, M., Hisano, K., Zhu, M. et al. (2019). Flexible multifunctional sensors for wearable and robotic applications. *Advanced Materials Technologies* 4: 1800626.

10 Bowen, C.R., Taylor, J., Leboulbar, E. et al. (2014). Pyroelectric materials and devices for energy harvesting applications. *Energy and Environmental Science* 7 (12): 3836–3856.

11 Bowen, C.R., Kim, H.A., Weaver, P.M., and Dunn, S. (2014). Piezoelectric and ferroelectric materials and structures for energy harvesting applications. *Energy and Environmental Science* 7 (1): 25–44.

12 Nurettin, S. and Muammer, K. (2021). A comprehensive review on the state-of-the-art of piezoelectric energy harvesting. *Nano Energy* 80: 105567.

13 Lendlein, A. and Trask, R.S. (2018). Multifunctional materials: concepts, function-structure relationships, knowledge-based design, translational materials. *Multifunctional Materials* 1: 010201.

14 Gibson, F.G. (2010). A review of recent research on mechanics of multifunctional composite materials and structures. *Composite Structures* 92: 2793–2810.

15 Haertling, G. (1999). Ferroelectric ceramics: history and technology. *Journal of the American Ceramics Society* 82 (4): 797–818.

2

Piezoelectric Fundamentals

This chapter introduces the *piezoelectric* effect in materials and describes an important subclass of piezoelectric materials, namely *ferroelectric* materials. The origin of the piezoelectric effect is explained, and example materials are discussed. This chapter also provides a description of the *pyroelectric* effect since many, but not all, piezoelectric materials are also pyroelectric. Efforts to further improve properties by the development of piezoelectric composites are also introduced. The characteristics of these classes of materials will be discussed in more detail in Chapter 3.

2.1 Piezoelectric Materials

Jacques and Pierre Curie first reported piezoelectricity in 1880, following a systematic investigation into the effect of charge generation by the application of force in crystals of Rochelle salt (sodium potassium tartrate), quartz and tourmaline [1]. Hankel proposed the now-accepted term for this phenomenon, derived from the Greek 'piezen' (to press) and electricity; hence, the term 'piezoelectricity' relates to the generation of electricity through the application of pressure. The contemporary definition offered by Cady is that piezoelectricity is '*electric polarisation produced by mechanical strain in crystals belonging to certain classes, is proportional to the magnitude and sign of the imposed strain*' [2].

Piezoelectric materials, therefore, develop an electrical charge when subjected to mechanical strain, and this is termed the *direct* piezoelectric effect. Piezoelectric materials will also develop a mechanical strain when subjected to an applied electric field, E_f, which is termed the *converse* piezoelectric effect. For these reasons, they are commonly used as both actuators (Chapter 4) and sensors (Chapter 5). Recently, there has also been increasing interest in the use of the direct piezoelectric effect to generate electrical energy by accumulating the charge generated; this is often termed 'energy harvesting' or 'energy scavenging' (Chapter 6) [3]. Common piezoelectric ceramics include quartz, barium titanate ($BaTiO_3$) and lead zirconate titanate ($Pb(Zr,Ti)O_3$ or PZT), and we will see that the materials can be produced in both single-crystal form and polycrystalline form in a variety of shapes including bulk materials, thin films and even as fibres in an effort to create composite systems. The PZT family of materials is the most commonly used due to its attractive piezoelectric

Piezoelectric Materials: Properties, Applications and Devices, First Edition.
Christopher R. Bowen and Andrew R. Plummer.

properties. Since its development in the 1950s, compositional improvement in PZT has led to relatively incremental increases in the piezoelectric performance, and more significant recent progress has been made in relation to single crystals, such as relaxor-$PbTiO_3$ [4]. Hence, improvements in piezoelectric device performance have often originated from a combination of the development of new materials and novel component configurations, which optimise actuation and sensing capabilities for a specific application.

The direct piezoelectric effect is shown schematically in Figures 2.1a,b, where the direction of polarisation (P) of the piezoelectric material is indicated by the macroscopic dipole within the material. The direct piezoelectric effect can be considered as the generation of a potential difference and an associated electric field across the opposite faces of a piezoelectric crystal when a mechanical stress is applied between these faces under open-circuit conditions. Under short-circuit conditions, a current will be generated. It should be noted that the direction of the electric field (open circuit) or current flow (short circuit) reverses as the direction of the applied stress is reversed, as in Figure 2.1a (compression) and Figure 2.1b (tension). This is the reason why piezoelectric materials often produce AC currents in sensing or harvesting applications.

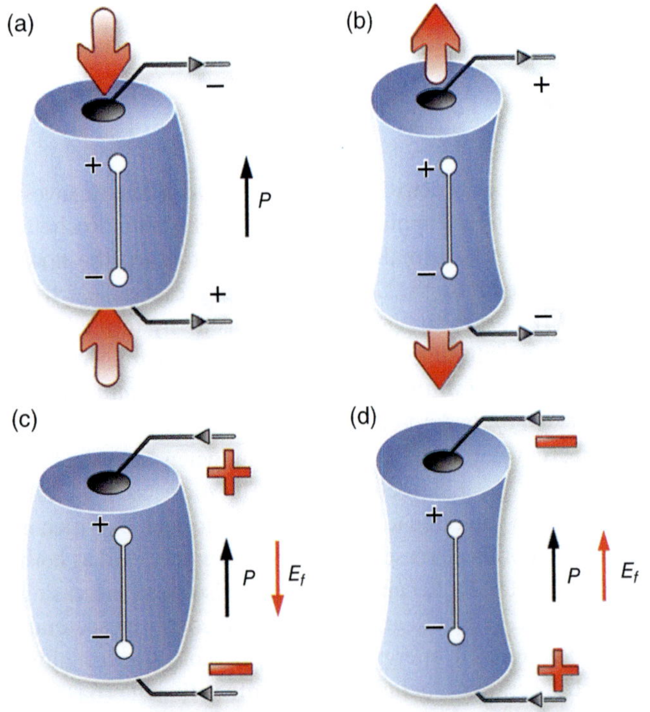

Figure 2.1 Simple schematic of piezoelectric behaviour. The *direct* effect: Application of (a) compressive or (b) tensile force generates a charge on the opposing faces and a potential difference at open circuit and current flow at short circuit. The *converse* effect: An applied electric field either (c) equal to or (d) opposite to the polarity of the material leads to the development of a strain. P is the polarisation direction, and E_f is the applied electric field. *Source:* Ref. [5] / with permission of Springer Nature.

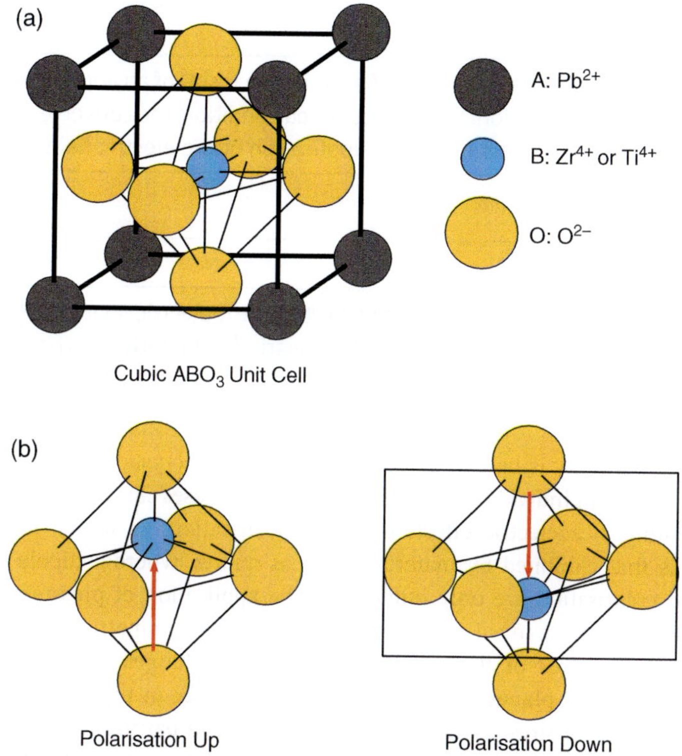

Figure 2.2 (a) Schematic of perovskite ABO_3 unit cell for PZT, at temperatures above the Curie point, it is a cubic perovskite structure; this unit cell is symmetrical, and there is no electrical dipole or polarisation. (b) At temperatures below the Curie point, the unit cell can be tetragonal, leading to a dipole, and the figure illustrates 180° polarisation reversal for two of the six possible polarisation states produced by the displacement of the central cation in the tetragonal phase. The distortion of the unit cell produces an electric dipole, e.g. the displacement of Ti^{4+} or Zr^{4+} upwards, which results in the unit cell having a net positive charge in the upper region of the cell. *Source:* Ref. [6] / with permission of John Wiley & Sons.

The converse piezoelectric effect is the change in strain for a free crystal (or stress for a clamped system) that accompanies the application of an electric field, as shown in Figures 2.1c and 2.1d. Again, the reversal of the direction of the applied electric field leads to the reversal of the piezoelectric strain. Figure 2.1 also defines the polarisation direction (P) of the piezoelectric material and the direction of the electric field (E_f) used throughout this book.

Under the appropriate conditions, the direct and converse effects in Figure 2.1 are reversible and correspond to sensor (direct) and actuator (converse) modes of operation, respectively.

Note 2.1

The direct piezoelectric effect is the generation of a potential difference across the opposite faces of a piezoelectric crystal when a mechanical stress is applied under open-circuit conditions. The direct effect will lead to a current flow under closed-circuit conditions.

Note 2.2

The converse piezoelectric effect is the development of a strain as a result of an applied electric field in the free, unclamped condition. In a fully clamped condition, the converse effect will lead to the development of stress under the application of the electric field.

Note 2.3

The direction of the electric field or strain generated by the piezoelectric material is dependent on the direction of the applied stress or the electric field relative to the polarisation direction.

The phenomenon of piezoelectricity is best described by considering a crystalline solid and its distribution of ions within an individual unit cell. Two different cases can exist that lead to the material exhibiting piezoelectric effects due to a polarisation, as shown in Figure 2.1. The first case is that exhibited by materials such as quartz, where a dipole moment and an associated polarisation are only induced by the application of pressure on the crystal; in the absence of mechanical pressure, the structure is symmetrical [7]. The second case is when the centres of the positive and negative charges are non-centrosymmetric within the unit cell planes when no pressure is applied; in this case, a polarisation is present even when no stress is applied. This is the case in barium titanate ($BaTiO_3$), a common piezoelectric material; we will see in the next section that this material is also *ferroelectric*.

The arrangements of ions are constrained to minimum energy positions, resulting in repeating spatial relationships that define the unit cell of the crystal lattice. All crystals may be divided into 32 crystallographic classes which depend on the geometry and symmetry of the unit cell present [6]. Twenty-one of these classes are non-centrosymmetric, of which all but one (due to other symmetry elements) exhibit piezoelectricity due to some form of polarisation that is present due to a lack of symmetry. It is important to note that it is the lack of symmetry in the distributions of positively and negatively charged ions in such crystalline materials that leads to the presence of an electric dipole, shown schematically in Figure 2.1, that ultimately leads to a direct and converse piezoelectric response.

2.2 Ferroelectric Materials

2.2.1 Non-centrosymmetric Unit Cells

In order to understand the operational use and limits of piezoelectric sensors and actuators, it is first of interest to understand the origin of the piezoelectric effect in ferroelectric materials such as barium titanate ($BaTiO_3$) and lead zirconate titanate (PZT). We will focus on these materials since they are widely used in piezoelectric sensors and actuators.

Ferroelectrics represent an important subgroup, or subclass, of piezoelectric materials, which are characterised by (i) the presence of a spontaneous polarisation in the unstrained, stress-free state, (ii) a capability to reorientate the direction of polarisation, for example, by reversing the orientation of the applied electric field and (iii) the polarisation is maintained on removal of the applied electric field.

The term 'ferroelectric' was introduced by Valasek [8] to emphasise the analogy with the behaviour of ferromagnetic materials, such as iron, where the direction of magnetisation could be reversed by changing the direction of an applied magnetic field; it should be noted that the physical origins of these phenomena are quite different. Ferroelectricity corresponds to certain crystal structures, of which the perovskite configuration is the most technologically significant since it is relevant to common piezoelectric materials such as $BaTiO_3$ and PZT; there is also growing interest in lead-free ferroelectrics due to environmental and health concerns of lead-based materials [9].

Ferroelectric materials are typically ionically bonded and therefore consist of positive and negative ions. If we consider PZT, which is a solid solution of $Pb(Zr,Ti)O_3$, at an elevated temperature, it exhibits a cubic crystal structure, which is shown in Figure 2.2a, and this crystal structure is symmetrical. The Pb^{2+} ions are located at the corners of the unit cell (termed the 'A-site'), and the Zr^{4+} or Ti^{4+} ions are located in the centre (termed the 'B-site'). It can be seen that the O^{2-} ions are located at the faces of the unit cell. In this state, the material unit cell is symmetric and exhibits no spontaneous polarisation. As a result, it is not ferroelectric and does not exhibit piezoelectric properties.

If the temperature is now reduced to below the Curie point (T_c), the material undergoes a phase change whereby the crystal structure changes from a cubic structure to a tetragonal structure. This phase change results in a distortion of the crystal lattice, whereby the positively charged Zr^{4+} or Ti^{4+} ions are now displaced from the centre of the unit cell. For example, in Figure 2.2b, the crystal structure has changed from a symmetric cubic structure to a tetragonal structure, leading to a positive centre ion (Zr^{4+} or Ti^{4+}) now being displaced upwards so that the upper region of the unit cell is more positively charged with respect to the bottom. The upward displacement of a positively charged ion creates a non-symmetrical unit cell with a net polarisation and a net electric dipole moment [4]. For this tetragonal system, there are six potential polarisation directions, since the central ion can displace upwards, downwards, left, right or in and out of the plane of the image in Figure 2.2. This displacement creates an asymmetric charge distribution around the midplane of the cell and hence a net electric dipole within the cell. The direction in which the central ion is displaced is termed the *polarisation direction* or *polar axis* (P). Due to the presence of an electric dipole, the application of an electric field along the polar axis of the unit cell will result in an extension or contraction of the unit cell, depending on whether the applied electric field is along, or opposite to, the polar axis of Figure 2.2b. This is the converse piezoelectric effect and is the basis of piezoelectric actuation since there is a contraction or expansion of the material as a result of the applied electric field, as shown in Figures 2.1c,d.

In addition to the tetragonal crystal structure, the PZT system can also exist in the rhombohedral form, which is also non-centrosymmetric. Figure 2.3 shows the B-site ion displacement and associated polarisation directions (P) for the tetragonal (Figure 2.3a) and rhombohedral (Figure 2.3b). For the rhombohedral phase, the Zr^{4+} or Ti^{4+} ion can move diagonally towards one of the A-site ions that form the outer cuboid. Hence, there are eight equivalent polar axes for the rhombohedral phase of the unit cell. We will see below, and in Chapter 3, that it is advantageous to use materials with a mixture of tetragonal

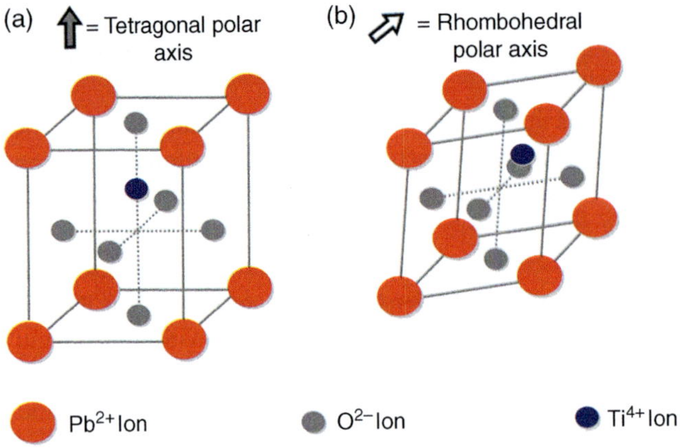

Figure 2.3 (a) Tetragonal perovskite (which exhibits six polarisation directions) and (b) rhombohedral perovskite (which exhibits eight polarisation directions).

(six polarisation directions) and rhombohedral (eight polarisation directions) phases in an attempt to maximise the number of polarisation directions and therefore the resulting piezoelectric properties.

If a ferroelectric sensor or actuator is heated above the Curie point, T_c, the transition from a non-centrosymmetric tetragonal or rhombohedral unit cell to a symmetrical cubic unit cell will clearly lead to the loss of ferroelectric (and piezoelectric) properties. Therefore, the T_c is strongly related to the upper operating temperature of the material as a piezoelectric transducer.

Note 2.4

A ferroelectric material is a material that has a spontaneous electric polarisation that can be reoriented by the application of an electric field, where the polarisation is retained when the electric field is removed.

Note 2.5

The piezoelectric effect in ferroelectric materials is the result of the non-centrosymmetric unit cell and the resulting dipole moment, which can change with stress.

Note 2.6

The T_c is a temperature that provides information regarding the *upper* operating temperature of a ferroelectric, since above this temperature the unit cell is symmetrical and no longer exhibits ferroelectric or piezoelectric effects. This will be discussed in more detail in Section 3.1.

2.2.2 Lead Zirconate Titanate ($Pb(Zr_xTi_{1-x})O_3$, PZT) Ferroelectrics

Chapter 3 will discuss a broader range of ferroelectric materials, but in this introductory chapter, it is of interest to consider the most popular commercial family, namely lead zirconate titanate; this is typically referred to as PZT. This material dominates the piezoelectric and ferroelectric ceramic materials due to its high electromechanical coupling and the ability to tailor the piezoelectric properties by the addition of appropriate dopant atoms. PZT is a solid solution formed between $PbTiO_3$ and $PbZrO_3$ and has the chemical formula $Pb(Zr_xTi_{1-x})O_3$, where $0 \leq x \leq 1$. A phase diagram for this ceramic system is presented in Figure 2.4.

As discussed, at temperatures above the T_c, it can be seen in Figure 2.4 that the unit cell is cubic and symmetrical and therefore exhibits no ferroelectric or piezoelectric properties. This temperature can range from 230 °C to 480 °C in Figure 2.4, depending on the exact composition of the $Pb(Zr_xTi_{1-x})O_3$ material. On cooling below the T_c, the structure undergoes a phase transformation to either the high-temperature ferroelectric rhombohedral ($F_{R[HT]}$) or ferroelectric tetragonal (F_T) phase, depending on the composition, i.e. whether it is $PbZrO_3$ rich (x approaching 1) or $PbTiO_3$ rich (x approaching 0), respectively. It can be seen in Figure 2.4 that a morphotropic phase boundary (*MPB*) exists between these two phases and occurs at the Zr:Ti ratio of 52:48. At this composition, the rhombohedral and tetragonal phases are present in similar proportions and the strongest piezoelectric activity is observed, leading to a larger number of opportunities for the polarisation to align with an applied electric field [6, 8]. Other phases present in the PZT system are a low-temperature rhombohedral phase ($F_{R[LT]}$) and an antiferroelectric orthorhombic phase (A_O).

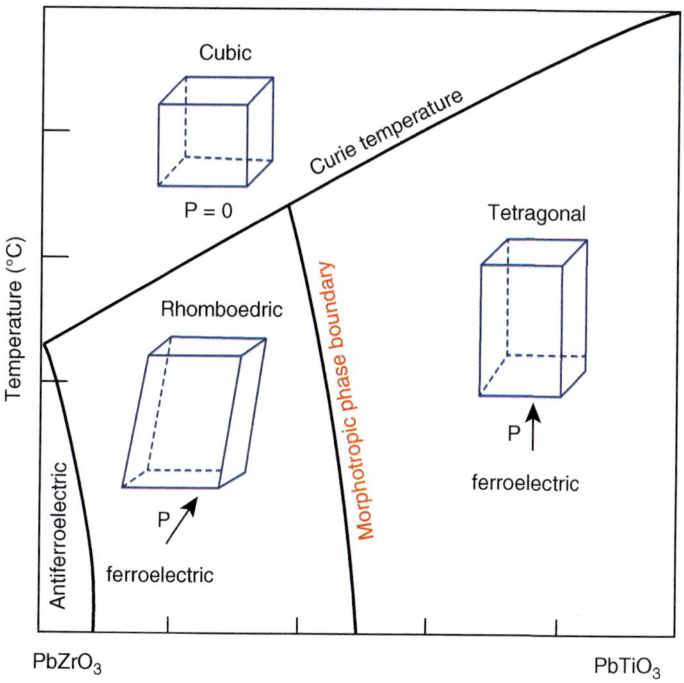

Figure 2.4 Phase diagram of the $Pb(Zr,Ti)O_3$ system showing the cubic, tetragonal and rhombohedral phases and morphotropic phase boundary (MPB). *Source:* Ref. [10] / Elsevier / CC BY 4.0.

Note 2.7

The displacement of the ions is restricted to specific crystal directions, allowing polarisation towards one of the six faces in a tetragonal cell or towards one of the eight corners for the rhombohedral phase. The most piezoelectrically active material compositions are often located at the morphotropic phase boundary, *MPB*, between these two phases.

2.2.3 Ferroelectric Domains

In a ferroelectric system, when the material is cooled below the T_c, the unit cells become spontaneously polarised in a range of directions depending on the crystal structure. While Figures 2.2 and 2.3 show the polarisation of a single unit cell and the generation of an electric dipole, regions in a ferroelectric material where the unit cells have equal spontaneous polarisation directions are known as *domains*. A twin structure is often generated, consisting of regions that have spontaneous polarisations with positive and negative polarity. Such a structure is shown schematically in Figure 2.5, and domain structures such as these form to reduce the net electrical potential and depolarising field across a crystalline solid.

Such a twin structure is known as the *domain structure* of the crystal. The twin component is called a *domain*, while the boundary between them is referred to as a *domain wall*. The domain walls in Figure 2.5 are termed 180° domain walls since the polarisations of the domains are orientated 180° to their neighbour.

In a tetragonal PZT crystal, the domain structure presented in Figure 2.5 can consist, for example, of regions of the 'polarisation up' and 'polarisation down' unit cells shown in Figure 2.2b [9]. In other regions of the crystal, the direction of spontaneous polarisation

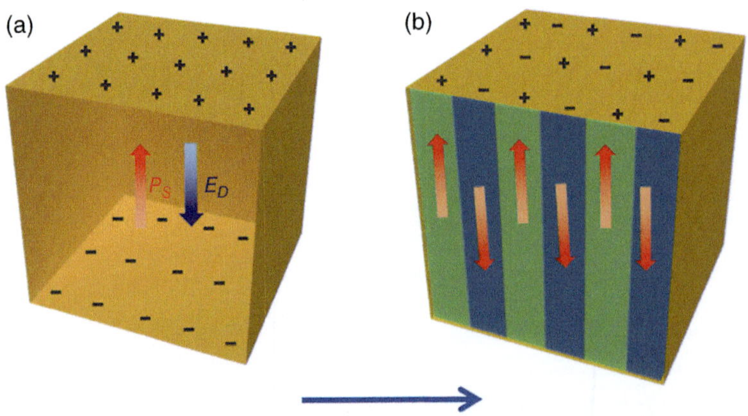

Figure 2.5 A simple domain structure with 180° domain walls. Arrows indicate the direction of spontaneous polarisation, whereby the individual unit cells of the material share a common polarisation direction (a) single domain showing polarisation direction and depolarisation field (b) formation of 180° domains to minimise depolarisation field. *Source:* Adapted from [10].

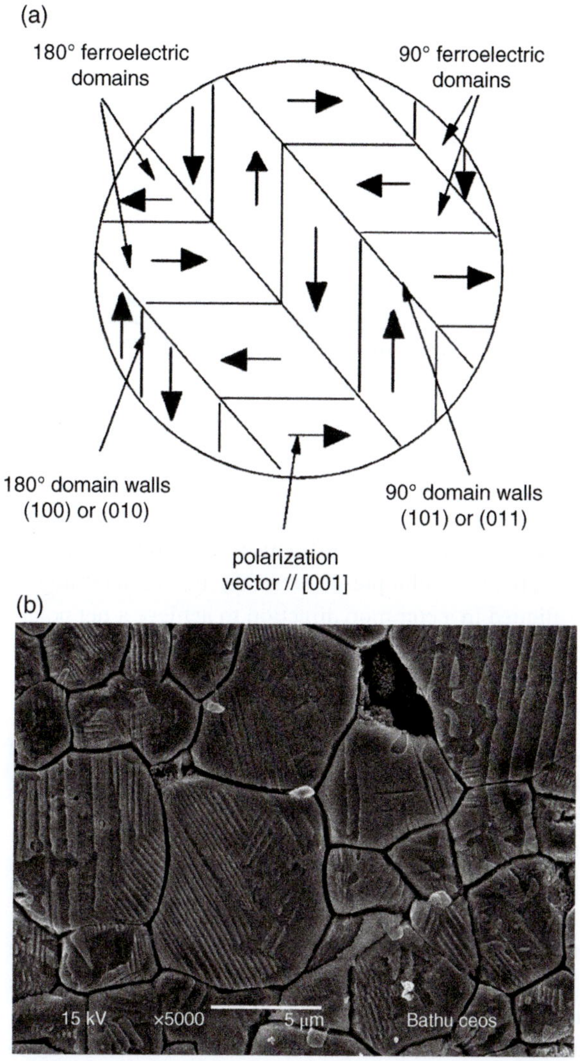

Figure 2.6 Domains in a tetragonal ferroelectric (a) Illustration of both 180° and 90° domains, showing the location of domain walls in the crystal. *Source:* Ref. [11] / with permission of AIP Publishing. (b) Scanning electron microscopy (SEM) image of ferroelectric ceramic showing grain and domain structure. *Source:* With permission of Rhodri Lewis.

may be in another direction allowed by the unit cell symmetry, such as the formation of 90° domain walls in tetragonal PZT, which can be seen in Figure 2.6a; 180° domain walls are also visible in Figure 2.6a. In rhombohedral compositions, in addition to the 180° domains, 71° and 109°orientations are possible, representing adjacent cell diagonals [12]. In polycrystalline materials, the domains are confined to individual grains, as in Figure 2.6b, which shows a scanning electron microscopy (SEM) image of a PZT material, where domains can be observed.

Note 2.8

Ferroelectric materials consist of domains, which are regions in a ferroelectric material where the unit cells have common polarisation directions. The domain structure is dependent on the crystal structure of the material, for example, tetragonal or rhombohedral, and therefore the potential polarisation directions.

2.2.4 Poling of Ferroelectric Materials

When a ferroelectric material is initially produced and cooled below the T_c, there is no net polarisation of the material since domains are randomly distributed over the volume of the material to reduce the net electric potential and minimise strain. The discovery of piezoelectricity in polycrystalline ferroelectric ceramics was a chance occurrence, following the initial investigations into the dielectric properties of ferroelectric barium titanate ($BaTiO_3$) [13]. In 1946, Gray discovered that the application of an electric field could reorientate and align the ferroelectric domains in $BaTiO_3$, thereby producing a net polarisation and piezoelectric behaviour in the polycrystalline material [6]. This is important since for a single-crystal or polycrystalline ferroelectric to exhibit piezoelectricity, the initial randomly orientated electrical dipoles must be aligned in a common direction to achieve a net polarisation. This is achieved in a ferroelectric material in a process termed *poling*. The process of alignment of ferroelectric domains (poling) is essential in converting a seemingly electrically inactive ferroelectric material into a useful electromechanically active material. Ferroelectric ceramics were rapidly adopted in preference to the earlier non-ferroelectric piezoelectric single-crystal materials, such as quartz, for several reasons: (i) They exhibit improved electromechanical performance, (ii) the polycrystalline ceramics were easy to fabricate, with an associated reduction in cost, and (iii) it was possible to freely choose the direction of polarisation by poling using an electric field, thereby allowing greater design freedom in device geometry [14]. This represents a significant advantage since for non-ferroelectric piezoelectric materials, such as quartz, which do not allow switching of the polarisation direction by the poling process, it is necessary to use single crystals that are cut in a specific orientation to achieve a piezoelectric response.

Poling involves the application of an electric field, usually at elevated temperatures to ease domain motion (but clearly not above the T_c). The poling electric field is used to orientate the polar axis of the ferroelectric domains to those directions allowed by the crystal symmetry, which are nearest to that of the applied electric field. The strong piezoelectric activity occurring at the MPB in PZT is thought to be associated with the larger number of polarisation directions available as a result of the presence of both tetragonal and rhombohedral phases (Figure 2.4).

Figure 2.7 shows a simple schematic of the poling process for a ferroelectric material. The initial material consists of randomly orientated domains, as shown in Figure 2.7a. The material is then heated to an elevated temperature below the T_c, to facilitate domain motion and switching, and a high electric field (typically a 1–2 kV/mm, but dependent on the material) is applied to orientate domains, such as 90° and 180° domains (Figure 2.7b). Unlike 180° domains, switching of 90° domains is a *ferroelastic* process and results in the material developing a permanent strain so that the material is slightly longer in the polarisation direction after poling. While the domains are fully aligned to the applied electric field in

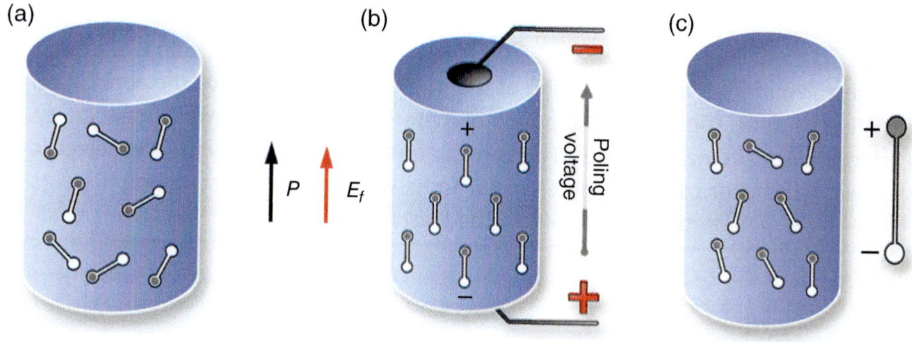

Figure 2.7 Poling of a ferroelectric ceramic. (a) After sintering, the ceramic has a notionally random arrangement of ferroelectric domains. (b) Application of an electric field at elevated temperature below T_c causes domains to align with the field. (c) Once cooled to room temperature, the field is removed, and remnant polarisation and strain are observed. The poled material has been rendered piezoelectric. *Source:* Adapted from [15].

the image in Figure 2.7b, it should be remembered that they are restricted to the polarisation directions, as in Figure 2.3, and benefit from using a composition at the MPB. After aligning the domains with the applied electric field, the material is then cooled to ambient temperatures while the electric field is still applied to ensure the domains remain aligned. On removal of the electric field at ambient temperatures, while some domains may change configuration (Figure 2.7c), the material retains a net polarisation, becoming a poled ferroelectric ceramic with a resultant piezoelectric behaviour. In addition to the orientation of 90° and 180° domains, reorientation of the 71° and 109° orientations is also possible.

Once poled, the ferroelectric has a polar axis, which for the case in Figure 2.7 is along the cylinder axis. The material is now piezoelectric and will exhibit the direct and converse piezoelectric effects; for example, the application of an electric field parallel to the poling direction (polar axis) results in an extension of the material, and the application of an electric field opposite to the polar axis results in contraction, as shown in Figure 2.1. An understanding of the poling process is important since it helps to understand the operational limits of a ferroelectric-based device; for example, during the operation of a piezoelectric, it is important to maintain alignment of the ferroelectric domains; otherwise, the piezoelectric properties are lost. A return to a random arrangement of domains, or a degradation of the degree of alignment, is possible if a piezoelectric actuator or sensor is subjected to

i) high service temperature, especially in relation to the T_c
ii) high mechanical loads (particularly compressive)
iii) high electric fields (especially when not applied parallel to the polar axis or against the polarisation direction)

Note 2.9

Ferroelectric materials are *poled* to align the ferroelectric domain. Loss of polarisation and degradation of piezoelectric properties can be due to exposure to high temperatures, high mechanical loads or incorrectly applied electric fields, all of which will lead to domain motion and loss of polarity.

Note 2.10

Ferroelectric materials allow switching of the polarisation direction by the application of an electric field. This allows the poling of materials in various device architectures to achieve piezoelectric properties. This is in contrast to non-ferroelectric piezoelectric materials, which are used in single-crystal form with specific crystallographic cuts relative to the polarisation direction(s) of the crystal.

2.3 Pyroelectric Materials

We have seen that piezoelectric materials can act as sensors of stress or strain due to the generation of a potential difference across the opposite faces of a piezoelectric crystal when a mechanical stress is applied under open-circuit conditions. We will see that this is related to the change in polarisation of the piezoelectric with the application of a stress. *Pyroelectric* materials are of interest as they can act as thermal sensors since their polarisation changes with temperature. These materials are also polar and have a polarisation P in the absence of an applied electric field, where P is the average electric dipole moment per unit volume.

A pyroelectric response originates from the dependence of polarisation level being dependent on the material's temperature. For example, when the material is heated ($dT/dt > 0$), its polarisation level decreases as dipoles lose their orientation due to thermal vibration, and the level of bound charge on the material surface decreases [16]; this results in an electric current under short-circuit conditions or an electric field under open-circuit conditions. Similarly, as the pyroelectric is cooled ($dT/dt < 0$), the dipoles regain their orientation, leading to an increase in the level of polarisation, thus reversing the electric current flow. It is worth noting that due to the polar nature of pyroelectrics, they are also piezoelectric, although not all piezoelectrics are pyroelectric; see Figure 2.8 [17]. For example, since quartz only exhibits a polarisation under an applied stress, the application of a temperature change to a stress-free quartz crystal will not lead to the development of

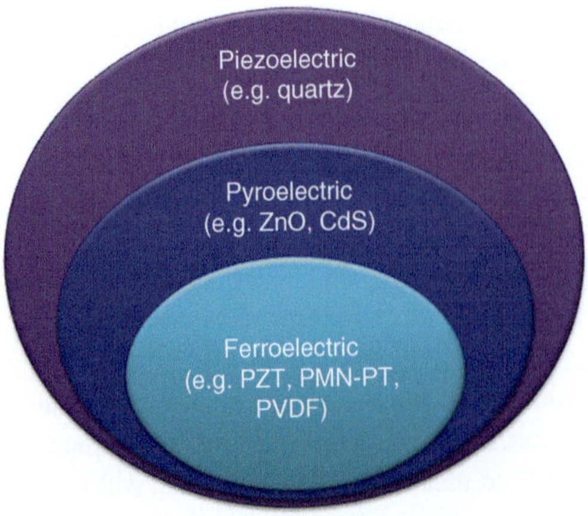

Figure 2.8 Piezoelectric, pyroelectric and ferroelectric material relationships. *Source:* Ref. [17] / With permission of Royal Society of Chemistry.

any pyroelectric charge. It can also be seen in Figure 2.8 that ferroelectric materials are a subclass of pyroelectric materials, which have a polarisation direction that can be switched (e.g. P to $-P$) by reversing the direction of the applied electric field. All ferroelectrics are therefore piezoelectric and pyroelectric once they have been poled. While this can be advantageous for multifunctionality, as described in Chapter 1, it can lead to complications in sensor applications where there is a combination of thermal and mechanical stimuli.

Note 2.11

All poled ferroelectric materials are pyroelectric and piezoelectric, since their polarisation changes with temperature and stress. All pyroelectrics are also piezoelectric, since all pyroelectrics are polarised. However, not all piezoelectric materials are pyroelectric or ferroelectric, since materials such as quartz only exhibit a polarisation as a result of an applied stress, and their polarisation direction cannot be switched by an electric field.

2.4 Piezoelectric Forms: Bulk, Thin Films and Fibre Composites

We have discussed that non-ferroelectric piezoelectrics are often in the form of single crystals, and this can be in the form of bulk crystals or thin films grown on substrates. Ferroelectric materials, in particular inorganic materials such as $BaTiO_3$ and PZT, are available in the form of polycrystalline ceramics, thin films or fibres. Single-crystal ferroelectrics are also available, some of which exhibit excellent properties [4], again in bulk or thin-film form.

One particular advantage of the ability to pole ferroelectrics is the possibility to produce a range of material architectures or composite structures to enhance specific properties, such as sensitivity, activity or mechanical flexibility. These are now discussed.

2.4.1 Piezoelectric Composites and Connectivity

A piezoelectric composite material generally consists of two material phases: an active piezoelectric phase and a passive (non-piezoelectric) phase. The connectivity of a piezoelectric composite determines its electrical, mechanical and thermal properties. Newnham et al. [18] defined the connectivity of a piezoelectric composite material with respect to the dimensionality of the interconnection of the individual phases. As a result, a two-phase material can be described by two numbers: the first of which refers to the piezoactive phase and the second to the passive phase. For example, a PZT with 3-0 connectivity would consist of a continuous active ceramic phase with an isolated second phase, such as porosity or a polymer. A 1-3 piezocomposite would consist of one-dimensional active piezoelectric rods or fibres that are embedded in a continuous polymer matrix, as shown in Figure 2.9a. Although originally defined for piezocomposites, this notation is now commonly used for general composite engineering. Figure 2.9b shows a selection of common two-phase structures that can be achieved [19].

While material developments continue, there is potential for further increases in performance using the piezoelectric composite approach. Current developments in piezoelectric

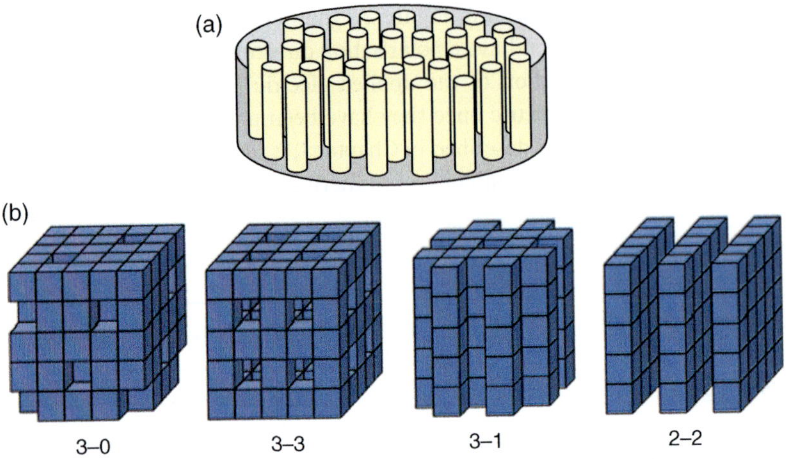

Figure 2.9 Example of composite architectures where the first number refers to the piezoelectric active phase and the second to the passive phase. (a) A 1-3 piezoelectric composite and (b) composites with 3-0, 3-3, 3-1 and 2-2 architectures. *Source:* [19] / with permission of IOP Publishing.

composites have focused on structural features and novel configurations to enhance the actuation and sensing capabilities of components. Successful devices have improved performance through mechanical amplification methods; these include the bimorph, Moonie, Thunder, Rainbow and a variety of other structures [20]. The development of piezoelectric composites, which typically combine active ferroelectric ceramics and passive polymers, provides new sensors or actuators that are combined to best exploit the advantages and minimise any deficiencies of the monolithic materials, such as sensing performance (to be discussed in Chapter 5), low toughness or low mechanical flexibility. Functional composites may be optimised by various means, such as the use of periodicity and scale to control resonant behaviour, connectivity patterns to create electric field and force concentrations and structural symmetry to alter mechanical properties.

Composites containing aligned individual PZT rods or fibres have been widely studied and utilised [21]. More recently, the availability of fine-scale PZT fibres (<300 μm in diameter) of semi-continuous lengths, micromachining and additive manufacturing has stimulated further development. Fine-scale PZT fibres were primarily developed for use in medical ultrasound transducers due to the improved imaging resolution provided by a reduction in the active fibre size [22]. Such fibres have also proved useful in structural composite applications, offering improved load transfer and dimensional compatibility for direct incorporation into a host structure. The inclusion of active fibres in this manner has attracted interest in the production of intelligent components capable of passive vibration control and damping [23].

2.4.2 Active Fibre Composites and Macro-Fibre Composites

As an example of a piezoelectric composite, the active fibre composite (AFC) and macro-fibre composite (MFC) can provide improved actuation through the use of interdigitated electrodes (IDEs) that apply an electric field in the plane of the composite and along the fibre length [24–26]; see Figure 2.10a. For a long array of piezoelectric fibres, or machined rods, embedded in a polymer matrix, an electrode can be applied to the composite either (i) across the fibre faces/fibre ends or (ii) along the fibre length using IDEs. For case (i), the fibre is poled along its length, and the electric field is parallel to the fibre length. However, the large electrode separation requires extremely high operating voltages since the electric field is voltage divided by electrode separation. For case (ii), when using IDEs, the operating voltage is much reduced since the electrode spacing is not reduced.

An additional advantage of the IDE approach is that, unlike case (i), the bending of the device does not significantly alter the electric field. In addition, damage to an individual fibre does not lead to a complete loss of strain capability, leading to increased damage tolerance. The disadvantage of the IDE structure is the relatively complex and uneven electric field now generated within the device; see Figure 2.10b. Directly beneath the electrodes, a 'dead zone' exists where there is no electric field and the material remains unpoled. Research on the effect of electrode geometry has shown that the key parameters that influence the strain output are electrode finger width, electrode finger separation, and substrate thickness. Between the electrode fingers, the electric field is relatively even in magnitude and direction, as in Figure 2.10b. This demonstrates that, in addition to material

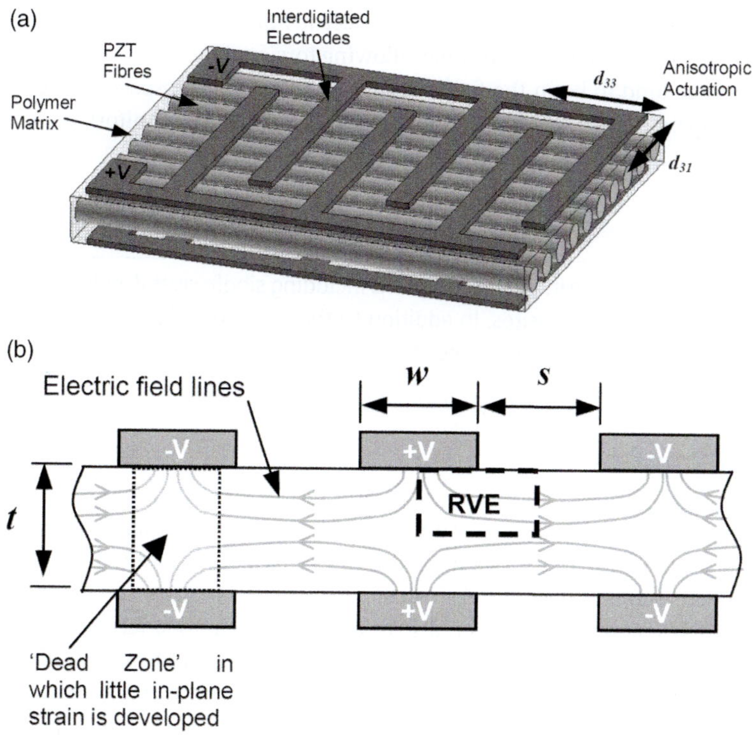

Figure 2.10 (a) Illustration of an active fibre composite, which utilises fine-scale fibres and interdigitated electrodes to provide superior actuation. (b) Electric field lines between electrodes; note the 'dead' zone beneath electrodes. *Source:* Ref. [10] / With permission of SPIE.

understanding, there is a need to understand the device configuration and associated electrical and mechanical fields experienced by the device during operation. Additive manufacturing is also providing additional opportunities to produce new device architectures [27].

Since AFCs combine fine-scale PZT fibres, embedded in a polymer matrix with IDEs, they provide the following advantages compared to conventional ceramic ferroelectrics:

i) Fine-scale fibres are sufficiently flexible to allow the actuator to conform to curved surfaces and complex shapes.
ii) Through the composite construction and the interdigitated electrode arrangement, damage tolerance is significantly improved when compared to bulk PZT material, which exhibits relatively low strength and toughness.
iii) Device flexibility and toughness allow the AFC to be readily bonded and attached to a variety of engineering structures.
iv) Large levels of piezoelectric strain are achieved since the IDE aligns the electric field along the direction of polarisation actuation; the larger d_{33} strain coefficient (a measure of strain per unit electric field) is used rather than the lower d_{31} strain coefficient; these coefficients will be discussed in more detail in Chapter 3.
v) The small electrode spacing of the interdigitated electrode leads to relatively low operating voltages when operated as an actuator. With a small electrode spacing, a large electric field can be achieved at low voltage.
vi) Since the composites consist of unidirectional piezoelectric fibres, the mechanical and piezoelectric coefficients are highly anisotropic, allowing torsional modes of actuation. Such fibre composites tend to be in the fields of shape control for aerospace applications, vibration suppression, energy harvesting and structural health monitoring.

Note 2.12

Piezoelectric materials are available in a variety of forms including single crystal, polycrystalline ceramics, thin films and composites. In addition to the selection of the optimum material, the use of composite structures enables further tailoring of the properties.

2.5 Concluding Remarks

As outlined in Figure 2.11, the investigation and use of piezoelectric materials have progressed significantly since the dawn of the twentieth century. Quartz (α-SiO_2) and Rochelle salt were initially utilised, and both were used in single-crystal form in early sound navigation and ranging (SONAR) applications. The discovery of polycrystalline ferroelectrics such as barium titanate ($BaTiO_3$, BT) and PZT was a major step forward mid-century, and composite architectures have been developed in recent decades to further improve properties. The latest developments in single-crystal ferroelectric relaxor-lead titanate show promise, with around three times the coupling between mechanical and electrical domains compared to commercial piezoelectric materials. The reader is referred to relevant reviews for a more detailed historical description of materials' developments [4, 6, 8].

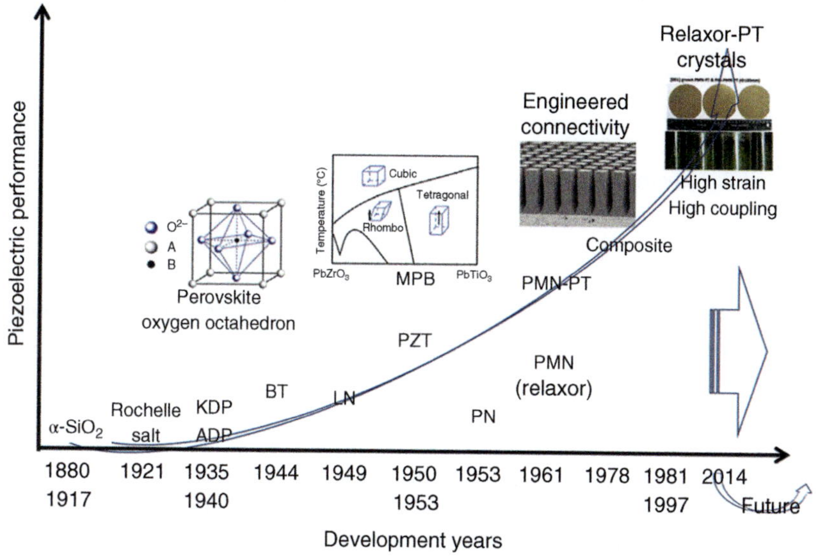

Figure 2.11 Milestone map for piezoelectric transducer material development. KDP/ADP: KH_2PO_4/$(NH_4)H_2PO_4$; BT: $BaTiO_3$; LN: $LiNbO_3$; PZT: $Pb(Zr,Ti)O_3$; PN: $PbNb_2O_6$; PMN: $Pb(Mg_{1/3}Nb_{2/3})O_3$; PMN–PT: $Pb(Mg_{1/3}Nb_{2/3})O_3$–$PbTiO_3$. *Source:* Ref. [4] / with permission of Elsevier.

This chapter has provided an initial introduction to piezoelectric, pyroelectric and ferroelectric materials and the origin of direct and converse piezoelectric effects in such materials. The poling process in ferroelectrics has been described, and the origin of pyroelectric behaviour has also been discussed. Examples of piezoelectric connectivity and piezoelectric composite structures have been introduced. The following chapter (Chapter 3) will describe such processes in more detail including mathematical relationships, including how to calculate the piezoelectric strain resulting from an applied electric field or the voltage generated by an applied stress. This will help to ensure operational limits of stress, temperature and electric field can be understood and materials and devices can be selected and used effectively.

References

1 Amato, I. (1997). *Stuff – The Materials The World Is Made Of.* New York: Basic Books.

2 Cady, W.G. (1964). *Piezoelectricity*, vol. Volume 1. New York: McGraw Hill.

3 Uchino, K. (2018). Piezoelectric energy harvesting systems—essentials to successful developments. *Energy Technology* 6: 829–884.

4 Zhang, S., Li, F., Jiang, X. et al. (2015). Advantages and challenges of relaxor-PbTiO3 ferroelectric crystals for electroacoustic transducers – a review. *Progress in Materials Science* 68: 1–66.

5 Baxter, F.R., Bowen, C.R., Turner, I.G., and Dent, A.C.E. (2010). Electrically active bioceramics: a review of interfacial responses. *Annals of Biomedical Engineering* 38 (6): 2079–2092.

6 Haertling, G.H. (1999). Ferroelectric ceramics: history and technology. *Journal of the American Ceramic Society* 82: 797–818.

7 Erhart, J. (2013). Experiments to demonstrate piezoelectric and pyroelectric effects. *Physics Education* 48: 438–447.

8 Cross, L.E. and Hardtl, K. (1992). Ferroelectrics. In: *Piezoelectricity* (ed. C. Rosen, B. Hiremath, and R. Newnham), 1–31. New York: American Institute of Physics.

9 Wu, J. (2018). *Advances in Lead-Free Piezoelectric Materials.* Springer.

10 Nelson, L., Bowen, C., Stevens, R., Cain, M., Steward, M. (2003), Modeling and measurement of piezoelectric fibers and interdigitaded electrodes for the optimization of piezofibre composites. *Proceedings Volume 5053, Smart Structures and Materials 2003: Active Materials: Behavior and Mechanics*, https://doi.org/10.1117/12.484738.

11 Floquet, N., Hector, J., and Gaucher, P. (1998). Correlation between structure, microstructure, and ferroelectric properties of integrated film: influence of the sol-gel process and the substrate. *Journal of Applied Physics* 84 (7): 3815–3826.

12 Ricote, J., Whatmore, R.W., and Barber, D.J. (2000). Studies of the ferroelectric domain configuration and polarization of rhombohedral PZT ceramics. *Journal of Physics- Condensed Matter* 12: 323–337.

13 Cady, W.G. (1964). *Piezoelectricity*, vol. Volume 2. New York: McGraw Hill.

14 Pohanka, R.C. and Smith, P.L. (1988). Recent advances in piezoelectric ceramics. In: *Electronic Ceramics – Properties, Devices and Applications* (ed. L.M. Levinson), 45–137. Marcel Dekker Inc: New York.

15 Dent, A. C. (2008). Novel active fibre composites. PhD thesis. University of Bath.

16 Lang, S. (2005). Pyroelectricity: from ancient curiosity to modern imaging tool. *Physics Today* 58: 831.

17 Bowen, C.R., Kim, H.A., Weaver, P.M., and Dunn, S. (2014). Piezoelectric and ferroelectric materials and structures for energy harvesting applications. *Energy and Environmental Science* 7 (1): 25–44.

18 Newnham, R.E., Skinner, D.P., and Cross, L.E. (1978). Connectivity and piezoelectric-pyroelectric composites. *Materials Research Bulletin* 13: 525–536.

19 Roscow, J.I., Zhang, Y., Kraśny, M.J. et al. (2018). Freeze cast porous barium titanate for enhanced piezoelectric energy harvesting. *Journal of Physics D: Applied Physics* 51: 225301.

20 Wang, L., Chen, W., Liu, J. et al. (2019). A review of recent studies on non-resonant piezoelectric actuators. *Mechanical Systems and Signal Processing* 133: 106254.

21 Tressler, J.F., Alkoy, S., Dogna, A., and Newman, R.E. (1999). Functional composites for sensors, actuators and transducers. *Composites Part A: Applied Science and Manufacturing* 30: 477–482.

22 Janas, V.F. and Safari, A. (1995). Overview of fine-scale piezoelectric ceramic/polymer composite processing. *Journal of the American Ceramic Society* 78: 2945–2955.

23 Yoshikawa, S., Selvaraj, U., Brooks, K.G., Kurtz, S.K. (1992). Piezoelectric PZT tubes and fibers for passive vibrational damping. In Safari, A., Liu, M., Kingon, A. and Hartling, G.H. (eds). *Proc. 8th IEEE Int' Symp. On Applications of Ferroelectrics*, Greenville, USA.

24 Bent, A.A. and Hagood, N.W. (1997). Piezoelectric fiber composites with interdigitate electrodes. *Journal of Intelligent Material Systems and Structures* 8: 903–919.

25 Hagood, N.W. and Bent A.A., (1999). Composites for structural control. Massachusetts Institute of Technology. Patent 687827 [US5869189].

26 Wilkie, W., Bryant, R., High, J. et al. (2000). Low-cost piezocomposite actuator for structural control applications. *In: Proceedings of SPIE Seventh Annual International Symposium on Smart Structures and Materials*, Newport Beach, USA.

27 Dawei, S., Yao, L., and Tomoaki, K. (2023). Review of the applications of 3D printing technology in the field of piezoelectric ceramics. *Resources Chemicals and Materials* 2 (2): 128–142.

3

Properties of Piezoelectric Materials

3.1 Introduction

It has been shown in Chapter 2 that piezoelectric materials are affected by both electrical and mechanical stimuli; hence, they are characterised by a range of electromechanical coefficients, which relate stress, strain, electric field and charge density. Linear relationships are usually used, and these are a good approximation, particularly at low amplitude, although, at high levels of stress and electric fields, there can be significant non-linearities; this is often a result of domain motion [1]. An important coefficient for actuation and sensing is the piezoelectric *d*-coefficient. For an actuator, which uses the converse piezoelectric effect, the *d*-coefficient represents the strain (S) per unit applied electric field (E_f) and has units of metres per volt (m/V):

$$S = d\,E_f \tag{3.1}$$

If the strain and field are measured in the same direction, this piezoelectric 'strain coefficient' can be directly used to determine actuator displacement per applied voltage.

In a sensor, which uses the direct piezoelectric effect, the *d*-coefficient represents the polarisation or charge density (D) generated per unit applied stress (σ), which has units of charge per unit force (C/N):

$$D = d\,\sigma \tag{3.2}$$

If the charge density and stress are measured in the same direction, this can be used directly to provide the charge generated per unit force. We will see later that sensors can also operate in a voltage mode, for example an electric field generated per unit stress, and this is related to the *g*-coefficient; see Eqs. (3.5) and (3.6).

A piezoelectric material will also exhibit conventional mechanical and electrical properties. Thus, the strain will also depend on the applied stress according to Hooke's law; therefore, Eq. (3.1) needs to be extended to include the case of non-zero stress:

$$S = s\,\sigma + d\,E_f \tag{3.3}$$

where s is the compliance of the material (the reciprocal of Young's modulus). Similarly, charge density also depends on the electric field according to the material permittivity (ε), and Eq. (3.2) needs to be extended; thus:

$$D = \varepsilon\,E_f + d\,\sigma \tag{3.4}$$

Piezoelectric Materials: Properties, Applications and Devices, First Edition.
Christopher R. Bowen and Andrew R. Plummer.

Equations (3.3) and (3.4) provide a complete description of linear behaviour when variables (stress, strain, field and charge density) act in one direction only.

Note 3.1

The piezoelectric d-coefficient (d_{ij}) provides the strain per unit electric field in actuator mode or the charge density per unit stress in sensor mode. The g-coefficient (g_{ij}) provides a measure of the electric field per unit applied stress for a sensor.

3.2 Constitutive Equations

3.2.1 Alternative Single-Axis Formulations

Equations (3.3) and (3.4) provide a complete description of the linear material behaviour with the values for compliance, permittivity and piezoelectric strain coefficient constant for the particular direction under consideration. Considering Eq. (3.3), the compliance (s) of a piezoelectric material can be measured by testing a material specimen with zero electric field (E_f), i.e. under short-circuit conditions; to denote this, we can use the notation: $s \equiv s^{E_f}$. Similarly, considering Eq. (3.4), the permittivity value (ε) is the measured constant of proportionality with zero stress (σ), i.e. under free displacement conditions; to denote this, we can use the notation: $\varepsilon \equiv \varepsilon^{\sigma}$. Likewise, the piezoelectric strain coefficient (d) can be measured using an equation with either zero stress or zero field, respectively.

If the compliance is measured in an open circuit, which means the charge density is zero instead of the field, a smaller value of compliance is obtained; this is termed s^D. In other words, the piezoelectric effect leads to a voltage being generated that strains the material that opposes the displacement, thereby increasing its effective stiffness. In this case, it is convenient to rearrange Eqs. (3.3) and (3.4) to form another pair of equations. From Eq. (3.4):

$$E_f = \frac{1}{\varepsilon^{\sigma}} D - g\,\sigma, \quad \text{where } g = \frac{d}{\varepsilon^{\sigma}} \tag{3.5}$$

Using Eq. (3.5) to substitute for E_f in Eq. (3.3) gives:

$$S = s^D \sigma + g\,D, \quad \text{where } s^D = s^{E_f} - \frac{d^2}{\varepsilon^{\sigma}} \tag{3.6}$$

An alternative piezoelectric coefficient, g, is used in this formulation and can be considered as a measure of the electric field generated per unit applied stress or the strain per unit applied charge density.

If the permittivity is measured in a blocked (i.e. clamped) configuration, which means the strain is zero instead of the stress, a smaller value of permittivity is obtained, ε^S. In other words, the stress generates an opposing charge, reducing its effective capacitance. In this case, it is convenient to rearrange Eqs. (3.3) and (3.4) to form the following pair of equations. From Eq. (3.3):

$$\sigma = \frac{1}{s^{E_f}} S - e\,E_f, \quad \text{where } e = \frac{d}{s^{E_f}} \tag{3.7}$$

And using this equation to substitute for σ in Eq. (3.4) gives:

$$D = \varepsilon^S E_f + e\,S, \quad \text{where } \varepsilon^S = \varepsilon^{\sigma} - \frac{d^2}{s^{E_f}} \tag{3.8}$$

An alternative piezoelectric coefficient, *e*, is used in this formulation, which can be considered as the charge density generated per unit applied strain or the mechanical stress per unit applied field.

Figure 3.1 illustrates the alternative mechanical and electrical constraints or boundary conditions. Mechanically, the material constraints are blocked (zero strain) or free (zero stress); electrically, the constraints are open circuit (zero charge density) or short circuit (zero electric field). These three different forms of the piezoelectric constant, *d*, *e* and *g*, are summarised in Table 3.1. Also included is another variant that is sometimes used, which is the electric field generated per strain, which is equivalent to the stress per charge density, denoted *h*.

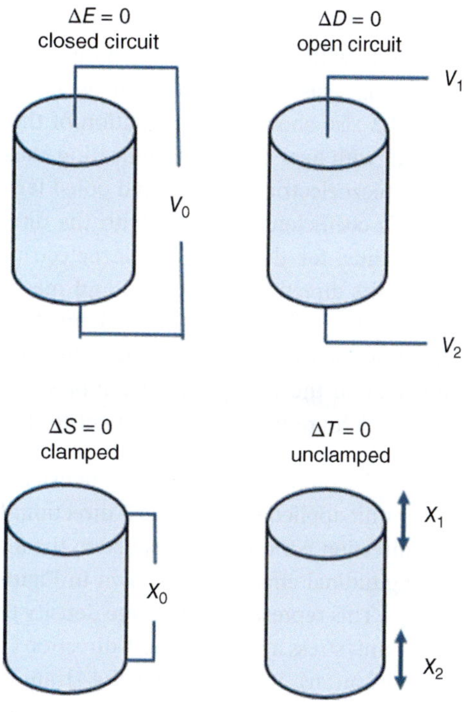

Figure 3.1 Electrical and mechanical constraint conditions for a piezoelectric. *Electrically constrained (top left to right):* Electrodes are short-circuited for a constant electric field (E_f) or open-circuited for constant charge density (*D*). *Mechanically constrained (bottom left to right):* The material is clamped (constant X_0) for constant strain (*S*) or mechanically free (variable X_1 or X_2) for a condition of constant stress (σ). *Source:* Adapted from Dent [2].

Table 3.1 Important piezoelectric coefficients *d*, *g*, *e* and *h* and their units.

Coefficient	Electrical output per mechanical input and SI units		Mechanical output per electrical input and SI units	
d	$\frac{\text{charge density developed}}{\text{applied mechanical stress}}$	C/N	$\frac{\text{strain developed}}{\text{applied electric field}}$	m/V
g	$\frac{\text{electric field developed}}{\text{applied mechanical stress}}$	Vm/N	$\frac{\text{strain developed}}{\text{applied charge density}}$	m^2/C
e	$\frac{\text{charge density developed}}{\text{applied strain}}$	C/m^2	$\frac{\text{mechanical stress developed}}{\text{applied electric field}}$	N/m/V
h	$\frac{\text{electric field developed}}{\text{applied strain}}$	V/m	$\frac{\text{mechanical stress developed}}{\text{applied charge density}}$	N/C

Note 3.2

For piezoelectric materials, it is important to (a) distinguish between compliance (or stiffness) measured in open-circuit or short-circuit conditions and (b) distinguish between permittivity measured in blocked or free displacement conditions.

3.2.2 Multi-Axis Linear Model

A piezoelectric crystal or a poled ferroelectric with highly aligned domains will exhibit anisotropic material properties; many of the important materials are often transversely isotropic. As a result, indices are used to indicate the particular direction of the material coefficients with respect to the polarisation direction, which are defined in Figure 3.2. The poling axis of the material is designated as the 3-direction (or z-direction). Figure 3.2 also shows the designation of the principal and shear axes for a piezoelectric material, with axes 4, 5 and 6 indicating shear around directions 1, 2 and 3, respectively.

Since piezoelectric materials and poled ferroelectric materials have a polar axis, the piezoelectric coefficients can vary with the direction of electrical and mechanical measurements. Thus, for the range of piezoelectric coefficients, subscripts i and j are used to indicate the direction of electrical and mechanical actions, respectively; for example, d_{ij}. Typically, the poling direction is defined as the 3-direction (or z-direction), while the 1-direction (or x-direction) and the 2-direction (or y-direction) are directions that are perpendicular to the poling axis. Examples of the d_{ij} coefficients, therefore, include those below, which are particularly relevant to the ferroelectric perovskites such as PZT:

i) d_{33} – This represents the charge density that is induced in the 3-direction (z-direction) per unit applied stress in the 3-direction. It can also denote the strain developed in the 3-direction per unit field applied in the 3-direction. This is often termed the primary or longitudinal effect and is shown in Figure 3.3 (left lower image).
ii) d_{31} – This represents the charge density that is induced in the 3-direction (z-direction), per unit stress applied in the 1-direction (x-direction, the direction perpendicular to the polarisation). It also denotes the strain developed in the 1-direction per unit electric field applied in the 3-direction. This is often termed the secondary or transverse effect and is shown in Figure 3.3 (middle lower image).

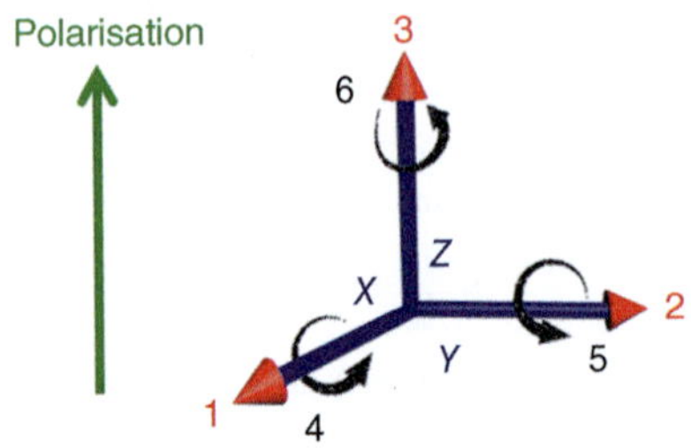

Figure 3.2 Designation of the principal (1, 2 and 3) and shear (4, 5 and 6) axes for a piezoelectric material. The direction of the poling electric field is indicated.

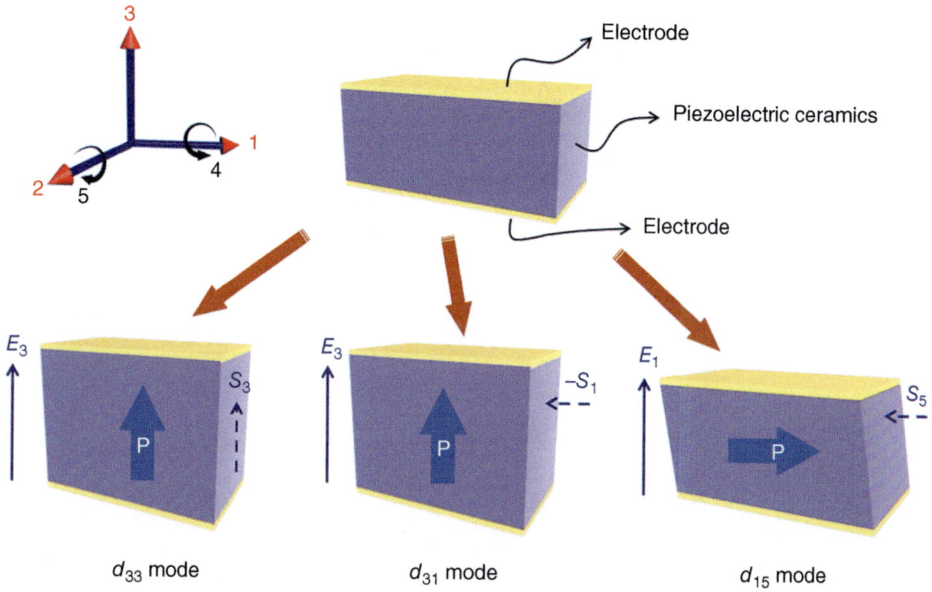

Figure 3.3 Modes of actuation. (left lower) d_{33}-mode with the field and displacement in the polarisation direction. (middle lower) d_{31}-mode with the field in the polarisation direction and displacement normal to the polarisation direction is used. (right lower) d_{15}-mode with the field applied being normal to the polarisation direction to shear the dipole.

iii) d_{15} – This is induced charge density in the 1-direction (x-direction, that is perpendicular to the polarisation direction) per unit shear stress that is applied in the 2-direction (perpendicular to both the 1- and polarisation directions). It also indicates the induced shear strain in the 2-direction per unit electric field applied in the 1-direction. The subscript 5 indicates that the shear is in the 2-direction, as shown in Figure 3.3 (right lower image).

A matrix equation is required to account for the possible interaction of stress, strain, electric field and charge in multiple directions. Thus, Eqs. (3.3) and (3.4) become:

$$\mathbf{S} = \mathbf{s}^{\mathbf{E_f}}\boldsymbol{\sigma} + \mathbf{d}^{\mathrm{T}}\,\mathbf{E_f} \tag{3.9}$$

$$\mathbf{D} = \boldsymbol{\varepsilon}^{\boldsymbol{\sigma}}\,\mathbf{E_f} + \mathbf{d}\,\boldsymbol{\sigma} \tag{3.10}$$

where σ and S are 6×1 vectors of stress and strain (the elements being in axial directions 1, 2 and 3 and shear directions 4, 5 and 6, respectively), and E_f and D are 3×1 vectors of electric field and charge density, respectively. The material properties are represented by s^{Ef}, a 6×6 compliance matrix; ε^{σ}, a 3×3 permittivity matrix; and d, a 3×6 matrix of piezoelectric coefficients.

Many piezoelectric materials, such as lead zirconate titanate (PZT) materials, are transversely isotropic, which results in some of the material property matrix elements being zero and others being equal to one another. Thus, writing Eqs. (3.9) and (3.10) in full, they often reduce to:

$$\begin{bmatrix} S_1 \\ S_2 \\ S_3 \\ S_4 \\ S_5 \\ S_6 \end{bmatrix} = \begin{bmatrix} s_{11} & s_{12} & s_{13} & 0 & 0 & 0 \\ s_{12} & s_{11} & s_{13} & 0 & 0 & 0 \\ s_{13} & s_{13} & s_{33} & 0 & 0 & 0 \\ 0 & 0 & 0 & s_{44} & 0 & 0 \\ 0 & 0 & 0 & 0 & s_{44} & 0 \\ 0 & 0 & 0 & 0 & 0 & 2(s_{11}-s_{12}) \end{bmatrix} \begin{bmatrix} \sigma_1 \\ \sigma_2 \\ \sigma_3 \\ \sigma_4 \\ \sigma_5 \\ \sigma_6 \end{bmatrix} + \begin{bmatrix} 0 & 0 & d_{31} \\ 0 & 0 & d_{31} \\ 0 & 0 & d_{33} \\ 0 & d_{15} & 0 \\ d_{15} & 0 & 0 \\ 0 & 0 & 0 \end{bmatrix} \begin{bmatrix} E_{f1} \\ E_{f2} \\ E_{f3} \end{bmatrix} \tag{3.11}$$

$$\begin{bmatrix} D_1 \\ D_2 \\ D_3 \end{bmatrix} = \begin{bmatrix} 0 & 0 & 0 & 0 & d_{15} & 0 \\ 0 & 0 & 0 & d_{15} & 0 & 0 \\ d_{31} & d_{31} & d_{33} & 0 & 0 & 0 \end{bmatrix} \begin{bmatrix} \sigma_1 \\ \sigma_2 \\ \sigma_3 \\ \sigma_4 \\ \sigma_5 \\ \sigma_6 \end{bmatrix} + \begin{bmatrix} \varepsilon_{11} & 0 & 0 \\ 0 & \varepsilon_{11} & 0 \\ 0 & 0 & \varepsilon_{33} \end{bmatrix} \begin{bmatrix} E_{f1} \\ E_{f2} \\ E_{f3} \end{bmatrix} \tag{3.12}$$

This gives a total of 10 coefficients to describe the material properties in all axes. We now consider a simple example of piezoelectric actuation.

3.2.2.1 Example Piezoelectric Element

Let us consider a free (unclamped) piezoelectric element 15 × 15 mm wide (w) and 1.3 mm thick (t), which is polarised through the thickness direction, as indicated in Figure 3.4. A potential difference of 1500 V is applied across its thickness using an electrode on the upper and lower surfaces of the element.

The material has a relative permittivity at constant stress in the 3-direction, namely the polarisation direction ($\varepsilon^{\sigma}_{r,33}$) of 2000 and piezoelectric coefficients $d_{33} = 500$ pC/N and $d_{31} = -240$ pC/N. The permittivity of free space (ε_0) is 8.85×10^{-12} F/m. Calculate the following.

i) Capacitance (C) of the piezoelectric element.
ii) Applied electric field (E_f).

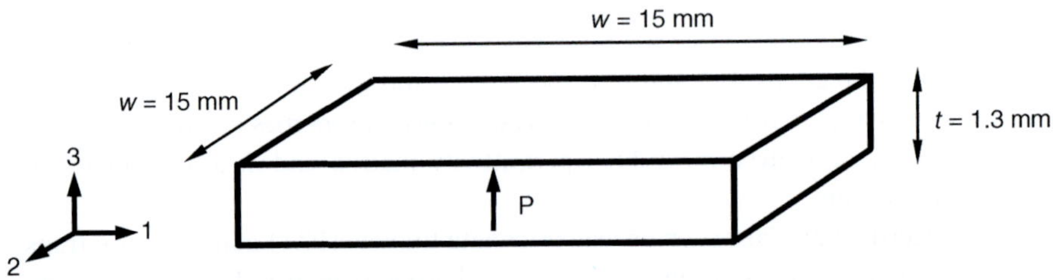

Figure 3.4 Piezoelectric material subject to an electric field. We assume that electrodes are located on the upper and lower surfaces.

iii) The resulting piezoelectric strain and displacement in the thickness direction, namely the 3-direction.
iv) The resulting piezoelectric strain and displacement in the width direction (1- or 2-direction).

Solution:

i) The capacitance (C) of a parallel plate capacitor with area (A) and thickness (t) is:

$$C = \frac{A.\varepsilon_{33}^{\sigma}}{t} \tag{3.13}$$

Since the relative permittivity of the material at constant stress is 2000, $\varepsilon_{33}^{\sigma} = 2000\varepsilon_0$. The permittivity at constant stress ($\varepsilon_{33}^{\sigma}$) is used in the calculation since the material is free and unclamped.

$$C = (0.015 \times 0.015 \times 2000 \times 8.85 \times 10^{-12})/(0.0013) = 3.06\ \text{nF}$$

ii) The applied electric field (E_f) for a specific potential difference (V) across a thickness (t) is:

$$E_f = \frac{V}{t} \tag{3.14}$$

Therefore, the applied electric field, $E_f = 1.15\ \text{kV/mm}$.

iii) The piezoelectric strain in the thickness (3-direction) is $d_{33}E_f$, which is:
strain in thickness $= d_{33}E_f = (500 \times 10^{-12}) \times (1.153 \times 10^{6}) = 0.00058$ (0.058%)
Since strain (S) is $\Delta t/t$, where Δt is the change in thickness, then

$$\Delta t = d_{33}E_f\, t$$

$$\Delta t = 0.75 \times 10^{-6}\ \text{m} = 0.75\ \mu\text{m}$$

It is of interest to note that since t is used to calculate both the electric field and displacement, it can also be calculated from $\Delta t = d_{33}\ V$.

iv) The piezoelectric strain in the width (1-direction) is $d_{31}E_f$, which is:
strain in width $= d_{31}E_f = (-240 \times 10^{-12}) \times (1.153 \times 10^{6}) = -0.00028$ (−0.028%).
Since longitudinal strain is $\Delta w/w$, where Δw is the change in width, then

$$\Delta w = d_{31}E_f w$$

$$\Delta w = -4.15\ \mu\text{m}$$

Here, while the thickness of the piezoelectric element increases, the width decreases.

Note 3.3
Piezoelectric strains are relatively small, where a maximum material strain of around 0.1% is typical.

Note 3.4

The d_{33} coefficients are usually larger than the d_{31} coefficients (typically $d_{33} \sim 2d_{31}$); however, a larger deflection can be achieved in the 1-direction compared to the 3-direction if the width is much greater than the height.

Note 3.5

The d_{33} and d_{31} coefficients are opposite in sign since, when the material expands in the polarisation direction (3-direction), it will contract in both the 1- and 2-directions. Many, but not all, piezoelectric materials are transversely isotropic, i.e. $d_{31} = d_{32}$.

3.2.3 Coupling Coefficients

In addition to piezoelectric coefficients such as d_{ij}, electromechanical coupling coefficients denoted k_{ij} describe the efficiency of energy conversion by the piezoelectric material from an electrical form to a mechanical form or vice versa. The ratio of the energy converted from one form, either electrical or mechanical, to a second kind (electrical or mechanical) can be defined as the square of the coupling coefficient (k_{ij}).

$$k_{ij}{}^2 = \frac{\textit{output mechanical energy}}{\textit{input electrical energy}} = \frac{\textit{output electrical energy}}{\textit{input mechanical energy}} \tag{3.15}$$

As outlined before, subscripts are again employed here in order to define the respective axes: the first describes the direction of the electric field or charge, and the second describes the direction of stress and/or strain.

For example, consider the application of a voltage to a piezoelectric element in the 3-direction, which is free to move (zero stress). The electrical input energy is half the product of the maximum voltage and charge, which in terms of field and charge density is:

$$\frac{1}{2} E_{f3} D_3 V_e \tag{3.16}$$

where V_e is the volume of the element. From Eqs. (3.3) and (3.4), or Eqs. (3.11) and (3.12), given the zero-stress condition:

$$D_3 = \varepsilon_{33}{}^{\sigma} E_{f3} \text{ and } S_3 = d_{33} E_{f3} \tag{3.17}$$

And so substituting into Eq. (3.16), the input electrical energy is equal to:

$$\frac{\varepsilon_{33}{}^{\sigma}}{2} \left(\frac{S_3}{d_{33}} \right)^2 V_e \tag{3.18}$$

Removing the voltage (resulting in zero field) in a mechanically blocked condition would recover some charge (consistent with ε_{33}^S, which is lower than ε_{33}^σ). Releasing the piezoelectric element would then allow the remaining energy to be recovered mechanically, and this is equal to half of the product of the maximum force and displacement, which in the context of stress and strain is:

$$\frac{1}{2}\sigma_3 S_3 V_e \tag{3.19}$$

From Eq. (3.3) or (3.11), given the zero-field condition:

$$S_3 = s_{33}{}^{E_f}\sigma_3 \tag{3.20}$$

And so substituting into Eq. (3.19), the recoverable (stored) mechanical energy is:

$$\frac{S_3{}^2}{2s_{33}{}^{E_f}}V_e \tag{3.21}$$

The ratio of the two energies, Eqs. (3.21) over (3.18), gives the square of the coupling coefficient:

$$k_{33}{}^2 = \frac{d_{33}{}^2}{\varepsilon_{33}{}^\sigma s_{33}{}^{E_f}} \tag{3.22}$$

The PZT-based piezoceramics typically have coupling coefficients (k_{ij}) in the range 0.7–0.75; therefore, the theoretical maximum energy conversion ($k_{ij}{}^2$) is approximately 50%.

Comparing Eq. (3.22) with the compliance and permittivity relationships in Eqs. (3.6) and (3.8), it can be seen that these can alternatively be expressed using the coupling coefficient:

$$s_{33}{}^D = s_{33}{}^{E_f}\left(1 - k_{33}{}^2\right) \tag{3.23}$$

$$\varepsilon_{33}{}^S = \varepsilon_{33}{}^\sigma\left(1 - k_{33}{}^2\right) \tag{3.24}$$

Similar expressions can be stated for other directions or combinations of directions. An interesting observation from Eq. (3.23) is that as the coupling coefficient $k_{33} \rightarrow 1$ (approaching 100% conversion), the open-circuit compliance $S_{33}^D \rightarrow 0$ and the material becomes infinitely stiff since all mechanical energy is converted to electrical energy. In addition, if $k_{33}{}^2$ is ~0.5 for a material like a PZT, it will be twice as stiff in an open-circuit condition compared to its closed-circuit condition. Similarly, the permittivity will be twice as large under a condition of constant stress compared to constant strain.

Note 3.6

The electromechanical coupling coefficient, k_{ij}, describes the efficiency of energy conversion by the piezoelectric material from electrical form to mechanical form, or vice versa. Any remaining energy is recoverable in the original form, i.e. the coupling coefficient does not imply any energy loss in the form of heat.

3.3 Polarisation–Electric Field Response of a Ferroelectric

For actuation applications, ferroelectrics such as PZT are typically subject to a high electric field (e.g. in the order of 1–3 kV/mm) [3, 4] in order to maximise the relatively small actuation strains we have seen in the example above. As a result, there is a need to measure (i) the commonly, and more easily, measured low-electric-field (< 1 V/mm) properties and (ii) the high-electric-field (> 1 kV/mm) properties. While the linear behaviour (Section 3.2) of a ferroelectric material is due to the response of the dielectric dipole to an electric field or mechanical stress, termed the *intrinsic response*, non-linear properties in ferroelectric materials are associated with the motion of domains in response to relatively high electric fields or stress, termed the *extrinsic response* [1], and this can lead to effects such as *hysteresis*.

In order to develop an understanding of how a ferroelectric material responds to the application of a high electric field, and overview of a polarisation–electric field (*P*–*E*) curve is necessary, where an example of a ferroelectric 'hysteresis' curve is presented in Figure 3.5.

The form of the polarisation–electric field hysteresis curve can be understood in terms of the response of the domain structure of a ferroelectric material to an applied electric field [3]. In its initial *unpoled* state (at point O), a large number of domains in the ferroelectric material are orientated randomly (since this is the lowest energy state); this initial random orientation of domains results in an overall polarisation of zero. We now consider the application of a relatively small electric field to the material, where a linear relationship between the level of polarisation and the applied electric field is observed since the material is acting like a simple linear dielectric material. This region can be seen in region O → A in Figure 3.5. If the magnitude of the applied electric field is now increased, the randomly orientated ferroelectric domains begin to align with the applied electric field. As a result of this domain 'switching' in the electric field direction, there is a resulting increase in the level of polarisation; this can be seen in region A → B in Figure 3.5. At a sufficiently high electric field, all of the ferroelectric domains will be aligned along the direction of the applied electric field; this can be defined as a state of *saturation*; see region B → C in

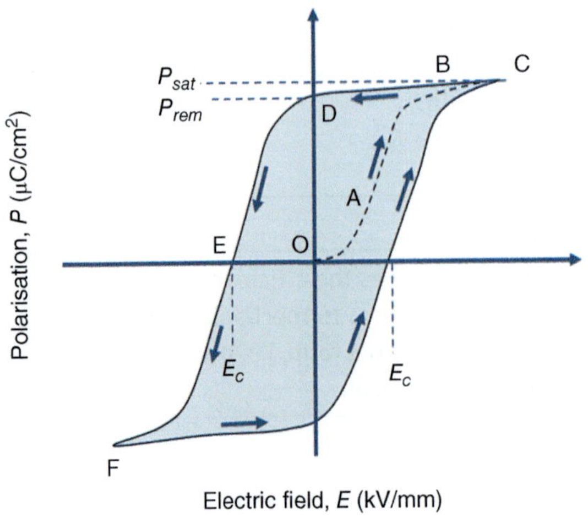

Figure 3.5 Diagram of the polarisation–electric field curve for a ferroelectric, also termed a *hysteresis* curve. *Source:* Adapted from Dent [2].

Figure 3.5. If the magnitude of the applied electric field is subsequently reduced, the level of polarisation will not return to zero in a ferroelectric but follow a path C → D in Figure 3.5. When the applied electric field is reduced to zero, a fraction of the ferroelectric domains that were aligned along the direction of the applied electric field at saturation will remain aligned. As a result, the material is considered to have a *remnant polarisation*; see P_{rem} (path O → D) in Figure 3.5. In this state, the material can exhibit piezoelectric properties and be used as a sensor or actuator since the remnant polarisation changes with stress (as a sensor), or the remnant polarisation changes with the field (as an actuator). This is analogous to the 'poled' state described in Chapter 2. It is the remnant polarisation, due to domain switching, that leads to the hysteresis curve in Figure 3.5.

To fully remove the remnant polarisation, P_{rem}, that is formed by the application and subsequent removal of a high electric field, there is a need to reverse the direction of the applied electric field direction. The magnitude of the electric field that is needed to reduce the level of polarisation back to zero (by following the path D → E in Figure 3.5) is defined as the *coercive electric field*, E_c. If the magnitude of the applied electric field is further increased in the negative direction, there will eventually be a condition where there will be alignment of all ferroelectric domains in the opposite direction (as indicated by the path E → F in Figure 3.5), leading to *saturation* but now in the opposite direction. The polarisation–electric field loop can be fully completed by now reversing the electric field direction to the original direction (as indicated by region F → C), which results in complete domain reversal. The somewhat tilted nature of the polarisation–electric field hysteresis loop, rather than a fully square loop, is a consequence of ferroelectric domains switching at a distribution of electric fields since there are a variety of domain types, which react differently to the applied electric field, for example 180°, 90°, 71° and 109° (see Chapter 2).

The nature and form of the polarisation–electric hysteresis loop are able to provide a good overview and understanding of the properties and behaviour of a ferroelectric material [2, 5] in response to an applied electric field and how the material will respond during operation, where it may be subject to both low and high electric fields. This includes [3]:

i) the remnant polarisation (P_r) of the ferroelectric can be measured; a high P_r indicates that the piezoceramic has a high internal polarisability,
ii) the coercive field (E_c) of the material can be estimated, which is the electric field necessary to remove polarisation or switch the direction of the domains; this can help understand the required poling conditions (field level) or the magnitude of the field that can be applied in a direction against the poling direction (e.g. for contracting an actuator) while avoiding domain switching,
iii) the squareness and shape of the loop can help understand the level of homogeneity and uniformity of the material; tilting of the polarisation–field loop is a consequence of ferroelectric domains switching at a range of electric field levels,
iv) the existence of a non-symmetrical polarisation–field loop or a loop that is not located at the centre of the polarisation–field axis indicates the existence of some form of ageing or the presence of internal space charge that leads to an internal bias that limits domain motion in a specific field direction,
v) the sharpness of the hysteresis loop provides an indication of the electric resistivity of the ferroelectric materials since conduction losses contribute to the measured polarisation–field loop, and

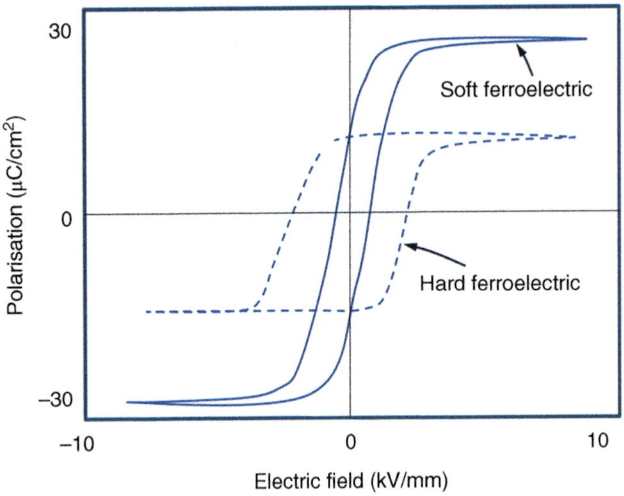

Figure 3.6 Hysteresis loops (polarisation–field loops) for a hard-type PZT and soft-type PZT. *Source:* Adapted from Steinhausen et al. [6].

vi) the slope of the polarisation–electric field along the hysteresis curve is related to the large-signal dielectric constant since $\frac{\delta P}{\delta V} = \frac{\delta Q}{\delta V \cdot A} = \frac{C}{A}$

Figure 3.6 shows polarisation–field loops for two different types of PZT, a 'hard' PZT where domain motion is more difficult compared to a 'soft' PZT. A hard PZT will exhibit a larger coercive field compared to a soft PZT since domain mobility is restricted. The soft PZT exhibits a higher remnant polarisation than the hard PZT as a result of the greater degree of domain switching for a specific applied electric field.

For an excellent discussion on 'decoding the fingerprint of ferroelectric loops', the reader is referred to [7]. In addition, for a discussion of common errors associated with interpreting polarisation–electric field loops, the reader is also referred to [8].

Note 3.7

The coercive field (E_c) indicates the electric field that is required to remove any polarisation and begin to switch the direction of ferroelectric domains.

Note 3.8

'Hard' PZTs have more restricted domain motion compared to 'soft' PZTs, leading to higher coercive fields. Applications related to hard and soft ferroelectric materials are described in Section 3.5.

3.4 Strain–Field Response of a Ferroelectric

In addition to forming a classical polarisation–electric field (*P–E*) hysteresis curve, the application of an electric field to a ferroelectric material also generates a characteristic strain–electric field (*S–E*) curve; this is often termed a 'butterfly' curve due to its shape. Figure 3.7 shows an example of a strain–electric field curve; it is informative to directly compare both the strain–electric field response in Figure 3.7 with the polarisation–electric field response in Figure 3.5.

The regions O → F in Figure 3.7 are the same as O → F in Figure 3.5; therefore, it is good to examine both figures together to consider the changes in both polarisation and strain with the applied field [3]. In Figure 3.7, in the initial *unpoled* state, the ferroelectric domains will be orientated in a random manner since it is the lowest energy state, and the ferroelectric is in its lowest state of strain (see point O in Figure 3.7). If we now apply a relatively small electric field to the ferroelectric (i.e. below the coercive field), the strain will not increase since the ferroelectric domains are randomly orientated and there will be no piezoelectric strain (this is indicated by region O → A in Figure 3.7). As the magnitude of the applied electric field is increased to point A, ferroelectric domains will begin to switch and orientate their polarisation direction to align in the direction of the applied electric field. While there will be a change in the level of polarisation in this region (Figure 3.5), there is also a dimensional change. If the magnitude of the applied electric field is now further increased, additional ferroelectric domains will begin to switch their polarisation direction; as a result, the piezoelectric strain will continue to increase with the applied electric field (this can be seen in region A → B in Figure 3.7). There will eventually be a condition, in terms of the magnitude of the electric field, whereby all of the ferroelectric domains are fully aligned with the applied field (termed *saturation*). In this condition, a further increase in field level will lead to a lower increase in the level of piezoelectric strain (B → C in Figure 3.7) since there is no longer any domain switching and only a piezoelectric response due to an intrinsic dipole extension. This is the same saturation condition shown in region

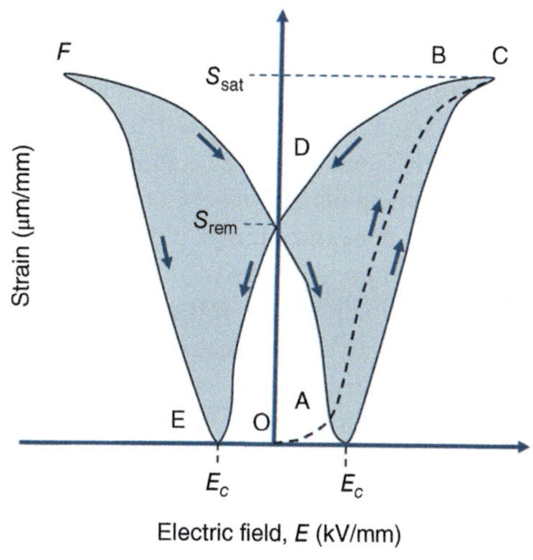

Figure 3.7 (a) Diagram of strain–electric field (*S–E*) curve of a ferroelectric material. This is sometimes termed a *butterfly* loop. *Source:* Adapted from Dent [2].

B → C in Figure 3.5 in terms of the polarisation–electric field response. At point C in Figure 3.7, the ferroelectric is considered to have reached a saturation strain, S_{sat}.

If we now remove the applied electric field, the strain of the ferroelectric will not return to zero; instead, it will follow the path C → D shown in Figure 3.7. At a condition of zero applied electric field, a fraction of the domains that were aligned by the applied field during saturation will remain aligned in the field direction. As a result, the material will exhibit a remnant strain, S_{rem} (see region C → D in Figure 3.7); this region is analogous to the formation of a remnant polarisation in region C → D in Figure 3.5. If the applied electric field is through the thickness of the material during the cycle O → D, the material is now thicker compared to its initial unpoled dimensions.

To remove the remnant strain that is produced by the poled ceramic, we must reverse the direction of the applied electric field. The magnitude of the electric field that is required to produce a strain level of zero (and return to its original state of strain) is termed the 'coercive field', E_c; see region O → E in Figure 3.7. This condition is analogous to the coercive field indicated in the polarisation–electric field loop in Figure 3.5. It is of interest to note that at the coercive field, point E in Figure 3.7, the strain is now the same as the initial unpoled material at point O since, at both of these points (O and E), there is no net polarisation and the ferroelectric domains can be considered to be effectively randomly organised.

If we now continue to increase the magnitude of the applied electric field in the negative direction, namely beyond point E, it will lead to a reversal of the direction of alignment of ferroelectric domains. This will then lead to an increase in the level of strain (see E → F in Figure 3.7), as domains are beginning to switch, and extend, in the direction of the applied electric field. The hysteresis cycle can be completed by now reversing the direction of the applied electric field (see region F → C in Figure 3.7). As a result of the characteristic shape of the strain–electric field loop of a ferroelectric, the strain–field is often described as a 'butterfly loop'.

The gradient of the strain–field loop in Figure 3.7 is a measure of the *d*-coefficient since we have seen that this can be defined as the piezoelectric strain induced per applied electric field; see Table 3.1. Examination of Figure 3.7 shows that the gradient, and therefore the piezoelectric *d*-coefficient, is zero at the coercive field (E_c, point E), and this makes sense as it is a result of there being no net polarisation at this point, as observed in Figure 3.5.

In terms of the operation of a poled piezoelectric actuator, it is important to avoid domain switching that can lead to depolarisation. Therefore, after the poling process (O → C), the actuator can be actuated and extended by the application of an electric field parallel to the poling direction, namely between points D and C. In an effort to increase the displacement and strain of a piezoelectric actuator, a degree of contraction can also be achieved by the application of an electric field that is in the opposite direction to the polarisation direction however. If the magnitude of the applied electric field inducing the contraction is too large, it can lead to domain switching and loss of polarisation at point E. The maximum applied electric field during contraction is therefore typically well below the coercive field, for example, $\sim 0.3E_c$. A possible extension–contraction cycle for a piezoelectric actuator is therefore an electric field range between $+E_{sat}$ and approximately $-0.3E_c$; there is likely to be some hysteresis, as a degree of domain motion would be expected in this wide electric field range.

3.5 Material Properties and Selection of Materials

To select the best piezoelectric material for a transducer, such as a sensor or an actuator, there is a need for an understanding of the range of materials which are available to the potential user or transducer designer. Research into the development of ferroelectric ceramics has led to the use of a wide range of additives, which are termed *dopants*. These additives are used to enhance material characteristics for specific applications [5, 9–11] and lead to a change in a wide range of properties, including the polarisation–field and strain–field responses shown in Figures 3.5 and 3.7. Isovalent, donor, acceptor and valence additives are employed to produce lead-based and lead-free-based ceramics and tailor specific properties for a range of sensor or actuator requirements.

Common ferroelectric ceramics that are used and available commercially include [3]:

i) barium titanate ($BaTiO_3$)
ii) lead zirconate titanate (PZT, $Pb[Zr,Ti]O_3$)
iii) lead lanthanum zirconate titanate (PLZT)
iv) lead magnesium niobate (PMN)
v) lead magnesium niobate–lead titanate (PMN–PT)

The PLZT-based ceramics are often employed in electro-optic applications and can be manufactured in the form of thin films that are optically transparent. Lead magnesium niobate is a well-known relaxor material; the reader is referred to key texts in this area if they wish to examine these materials in more detail [12]. $BaTiO_3$- and PZT-based piezoceramics are both used more widely, and their properties are now overviewed.

3.5.1 Barium Titanate ($BaTiO_3$)

Barium titanate was the first ferroelectric polycrystalline material that was used to produce a piezoelectric transducer in the 1940s [9]. The material is used in capacitor applications due to its high permittivity and also in piezoelectric applications; typical characteristics are:

i) a low Curie point, T_C (120 °C),
ii) a relatively low electromechanical coupling factor (k_{ij}) compared to alternative piezoelectric ceramics, such as PZT.

There has been significant research and development in an attempt to modify the properties of $BaTiO_3$ to optimise their performance for specific applications. This has involved substituting Pb, Sr or Ca for Ba and Zn or Sn for Ti. Examples of the range of additives employed for $BaTiO_3$-based ceramics are outlined in Table 3.2 [3].

Table 3.2 Role of additives for $BaTiO_3$ compounds.

Sr^{2+}	Decreases the Curie point, T_c
Pb^{2+}	Increases the Curie point, T_c
Ca^{2+}	Increases the temperature range that the tetragonal phase is stable
Co^{2+}	Decreases electrical losses at high electric field

From Table 3.2, it can be seen that $SrTiO_3$, $CaZrO_3$ and $PbTiO_3$ affect the performance of the ceramic regarding its Curie point, T_c [9]. While $BaTiO_3$ is a piezoelectric material of interest for sensors and actuators when compared to the PZT family of materials, it has (i) a lower Curie point, T_c ~120 °C, (ii) a lower electromechanical coupling coefficient (k_{ij}) and (iii) a smaller relative permittivity (dielectric constant); (iv) it is also more difficult to polarise and pole [5]. Berlincourt has also reported that $BaTiO_3$ has a lower ferroelectric response compared to PZT [5]. The material is clearly lead free, which has led to renewed interest in $BaTiO_3$-based materials and their development.

3.5.2 Lead Zirconate Titanate (PZT, $Pb(Zr,Ti)O_3$)

The PZT family of materials has become the material of choice for many piezoelectric applications. These materials are often used in conjunction with dopants in order to produce a piezoelectric material that is optimised for specific applications. A wide variety of PZT types are available commercially [9, 11]. Isovalent, donor and acceptor additives have been employed as PZT dopants, and Table 3.3 provides a list of the range of typical additives used in lead-based PZT piezoelectric materials. For example, Sr has the same valency as Pb (+2) and is therefore termed an *isovalent* additive. Nb substitution, due to the ionic radius, would be with Ti at the B-site (see Figure 2.2 for the location of B-sites), and the valency for Nb is +5, while the valency of Ti is +4; thus, Nb is termed a *donor* additive. In contrast, for the substitution of Fe, which has a valency of +3, compared to Ti (+4), Fe is therefore termed an *acceptor* additive [5, 13].

Table 3.4 illustrates the effect of a range of isovalent additives used with PZT ceramics, where examples of the range of additives are listed in Table 3.5 [3].

An isovalent additive will affect the structure of the ceramic, for example replacing some Pb with Sr will lead to $Pb_{0.95}Sr_{0.05}Zr_{0.53}Ti_{0.47}O_3$ when the Sr additive is able to accommodate 5% of the A-sites [11] (see Figure 2.2 for location of A-sites). The effect of Sr and Ba on the

Table 3.3 Typical additives employed for lead zirconate titanate (PZT) ferroelectric ceramics.

Ag, Ba, Bi, Ca, Cr, Co, Fe, Hf, K, La, Mg, Mn, Na, Nb, Nd, Ni, Sb, Sn, Sr, Ta, U, W, Y, Zn

Table 3.4 Effects of isovalent additives in PZT ceramics.

1) Decrease in Curie point, T_c
2) Increase in permittivity
3) Small change in the electrical and mechanical quality factor (Q_f)
4) Increased temperature dependence

Table 3.5 Examples of isovalent additives for PZT compounds.

Ba^{2+} for Pb^{2+}
Sr^{2+} for Pb^{2+}
Sn^{4+} for Ti^{4+}
Sn^{4+} for Zr^{4+}

Curie point (T_c) of $PbZrO_3$ is observed by examination of the phase diagram, where there is a decrease in T_c, which indicates the temperature where there is a transition from the tetragonal (or rhombohedral) phase to the paraelectric (non-ferroelectric) and symmetrical cubic phase.

Substitutions that use donor additives lead to the development of vacancies within the crystallographic structure of the ceramic that can facilitate the motion and switching of ferroelectric domains. The introduction of vacancies therefore leads to the development of a high piezoelectric strain due to increased extrinsic contributions to the piezoelectric strain due to domain motion during actuation. However, this increase in strain can be at the cost of some decrease in the dielectric and mechanical properties. The type of ceramic that contains donor additives is often termed 'soft' PZT since domain motion is enhanced; there is some analogy in the terminology with 'soft' ferromagnetic materials here. Donor additives are able to produce vacancies in both the A- and B-sites of the crystal structure, such as replacing the Pb (2+) with La (3+), where the composition would be $Pb_{0.97}La_{0.02}Zr_{0.53}Ti_{0.47}O_3$, and in this case, 2% of the A-sites contain La (3+), thereby resulting in 1% of sites being vacant [11]. A range of common donor additives is shown in Table 3.6, where the effects of using the donor additives are outlined in Table 3.7 [3].

With the use of acceptor additives, different effects are observed compared to those observed with donor additives, where acceptor additives restrict domain motion. This leads to a reduction in any loss or hysteresis, especially in conditions with the application of high electric fields or mechanical stress. The use of acceptor additives enables the formation of vacancies at the O-sites of the crystal structure; for instance, the composition $Pb_{0.99}K_{0.01}Zr_{0.53}Ti_{0.47}O_{2.995}$ has the K (+1 valence) acceptor additive at 1% of the A-sites, and 0.5% of the O-sites are therefore vacant [11] (see Figure 2.2 for the location of O-sites). The solubility limit of any acceptor dopant is low, but the introduction of vacancies leads to limited domain reorientation, thereby producing small hysteresis loops (making them good for high-drive applications) and high coercive fields [9]. Examples of common acceptor

Table 3.6 Examples of donor additives employed in lead zirconate titanate (PZT) compounds.

La^{3+} for Pb^{2+}
Bi^{3+} for Pb^{2+}
Sb^{5+} for Zr^{4+}
Nb^{5+} for Zr^{4+}

Table 3.7 Example effects of donor additives in lead zirconate titanate (PZT) ferroelectric ceramics.

1) Small variance in T_c
2) Increase in permittivity
3) Increase in the piezoelectric coupling factor (k_{ij})
4) The electrical and mechanical quality factor, Q_f, decreases significantly, a measure of the sharpness at resonance
5) Reduced coercive field and greater ease of poling
6) More easily depolarised by applied mechanical stresses

Table 3.8 Examples of acceptor additives used in lead zirconate titanate (PZT) compounds.

K^{1+} for Pb^{2+}
Fe^{3+} for Zr^{4+} and Ti^{4+}

Table 3.9 Effects of acceptor additives in lead zirconate titanate (PZT) ceramics.

1) Small variance in Curie point, T_c
2) Decrease in permittivity
3) Decrease in the piezoelectric coupling factor
4) Electrical and mechanical quality factor, Q_f, significantly increases, which is good for resonance
5) Higher coercive field and therefore more difficult to pole
6) Lower compliance and therefore higher stiffness

Table 3.10 Summary of lead zirconate titanate (PZT) compositions.

PZT-4 doped with Fe acceptor additive
PZT-5 doped with Nb donor additive
PZT-6 doped with Cr valence/acceptor additive
PZT-7 doped with La donor additive
PZT-8 doped with Fe acceptor additive

additives and their effects on material properties are outlined in Tables 3.8 and 3.9, respectively. The acceptor-doped ferroelectric ceramics are often termed 'hard' PZTs since domain motion is restricted; again, there is some analogy here to hard ferromagnetic materials [3].

Valence additives, such as Cr, U and Mn, are described in this way since their effective valence can change with both time and temperature [11].

The wide variety of available PZT types and compositions is outlined in Table 3.10, which also outlines the range of additives/dopants that are employed.

Specific groups and types of PZT-based and $BaTiO_3$-based ceramics are defined in relation to their material properties and ultimately their performance; examples include the piezoelectric coefficients, permittivity, polarisation–field (P–E) response, strain–field (S–E) response and their Curie point (T_c). As a result, in addition to referring to a ferroelectric as a 'soft' or 'hard' ferroelectric, piezoelectric ceramics are grouped into 'Navy types', or European naval classifications. Details of the US military standards are outlined in Table 3.11, where the defence and naval applications are closely linked to active and passive sound navigation and ranging (sonar) applications since the piezoelectric is employed as a transmitter (actuator) or a receiver (sensor).

Materials such as PZT-4, PZT-5 and PZT-8 may be known as Navy types I, II and III, respectively [15]. European standards are somewhat similar to the classification scheme defined by the Navy types but are indicated as Type 100–800, rather than Type 1–8. Table 3.12 summarises some of the main properties of piezoelectric ceramics, which are defined using the European standards (Type 100–800).

Table 3.11 Naval type standard classifications, taken from DOD-STD-1376A.

Type I	This is a modified lead zirconate titanate generally recommended for medium-to-high-power acoustic applications. A resistance to depolarisation at high electric fields and/or mechanical stress makes them suitable for deep-submersion applications.
Type II	A lead zirconate titanate modified to yield higher charge sensitivity but less suitable for driving at high electric fields due to dielectric heating. These materials are suitable for passive devices, such as hydrophones. Good time stability.
Type III	This is similar to Type I, but the material has been improved for use at high electric drive due to lower losses. Its electric field dependency on dielectric and mechanical losses is substantially reduced. A good choice because of greater electromechanical activity.
Type IV	A modified barium titanate for moderate-electric-drive applications. This material exhibits a lower piezoelectric activity and lower Curie point (T_c) compared to lead zirconate titanate compositions.
Type V	A composition intermediate to Types II and VI.
Type VI	Similar to Type II with higher charge sensitivity and dielectric constant, at the expense of a reduced Curie point (T_c).

Source: Adapted from DOD-STD-1376A [14].

Note 3.9

A wide variety of ferroelectric types are available depending on operational requirements such as high Curie point, high- or moderate-drive (electric field) levels, the need for low dielectric losses, high/low coercive field, quality factor (sharpness of resonance) and high sensitivity or high strain. This is achieved with a variety of doping strategies, which include isovalent, donor, acceptor and valence additives.

Table 3.12 summarises in more detail the properties of a range of materials at low-electric-field and high-electric-field properties for some common grades of PZT, ranging from a 'soft' PZT, where there is easy domain motion, to a 'hard' PZT, where domain motion is difficult. Since domain motion is restricted, the hard ferroelectric materials exhibit a high coercive field, a high depolarisation stress and a high quality factor compared to the soft ferroelectric materials; the quality factor (Q_f) is a measure of the sharpness at resonance and is important for resonant-based devices, such as a transmitter operating at a specific resonant frequency. Hard PZT materials are often employed for high-drive (high electric field or stress) applications, such as transducers for ultrasonic baths or sonar emitters, since the material can be driven at high electric fields and frequencies without excessive heat generation and degradation due to hysteresis from domain motion.

Table 3.13 provides a selection of hard and soft PZT-based materials. Hysteresis losses can cause self-heating of the material, especially if high electric fields are applied at high frequencies (see Section 3.12).

As described for the 'soft' materials in Table 3.13, dopants are added to facilitate domain wall motion. These materials possess high d_{ij} coefficients, higher strain, higher remnant polarisation and high permittivity and so are more suited for sensor applications or

Table 3.12 General properties for piezoceramics measured on a disc specified by BS 50324-1:2002.

Property	Symbol	Unit	Type 100		Type 200		Type 300		Type 400		Type 500		Type 600		Type 700		Type 800	
			Hard PZT		Soft PZT		Very hard PZT		Barium titanate		Hard PZT		Very soft PZT		Lead titanate		Lead metaniobate	
			Min.	Max.	Min.	Max.	Min.	Max.	Min.	Max.	Min.	Max.	Min.	Max.	Min.	Max.	Min.	Max.
Free relative permittivity	$\varepsilon_{33}^{\sigma}$		1100	1600	1600	2500	800	1150	700	1400	300	850	2500		150	300	200	300
Dielectric loss factor	$\tan \delta_d$	10^{-3}		6		25		5		10		5		30		30		10
Planar piezoelectric coupling factor	k_p		0.55		0.55		0.5		0.23		0.4		0.55			0.1		0.1
Curie temperature	T_c	°C	300		330		300		100		250		180		230		400	
Mechanical quality factor	Q_m		300			100	800		400		800			100	500			20
Piezoelectric charge coefficients	d_{33}	10^{-12} C/N	250		400		200		100		150		500		40		70	

Table 3.13 Comparison of four grades of PZT.

		Soft →			→ Hard
	Material property	**PZT-5H**	**PZT-5A**	**PZT-4S**	**PZT-8M**
Low electric field	Relative dielectric constant, ε_{33}^{T}	3350	1770	1500	800
	Induced strain, d_{33} (pm/V)	555	392	360	215
	Coupling coefficient, k_{33}	0.790	0.774	0.861	0.767
	Young's modulus, Y (GPa)	48	53	60	77
	Depolarisation stress (MPa)	44	72	>150	>150
High electric field	Coercive field, E_c (kV/cm)	7.2	11.2	8	8.4
	Remnant polarisation (mC/m^2)	300	375	110	80
	Saturation strain (ppm)	3200	4100	2100	1400
	Rep. cycle strain (ppm)	2350	2800	1875	1070
	Rep. cycle hysteresis (%)	24	24	7	4
	Rep. cycle energy density (J/kg)	17.7	27.7	14.1	5.9

The high-electric-field properties are related to a cyclic electric field of $\pm 2 \times E_c$ for PZT-5H and PZT-5A and $\pm$ 48 kV/cm for PZT-4S and PZT-8M. Representative cycle (Rep. cycle) data are at an electric field of -75% E_c against the poling direction, to $+200\%$ E_c (or 48 kV/cm for PZT-4S and PZT-8M).
Source: Bent [16] / Massachusetts Institute of Technology.

low-frequency actuators. However, the soft compositions exhibit lower depolarisation stress, higher hysteresis, lower coercive fields and lower quality factors. The high losses and hysteresis of soft materials due to domain wall motion limit the use of high-frequency actuators and transducers.

Materials such as the PZT-5A in Table 3.13 exhibit good material properties for applications related to actuators that are driven at relatively high electric fields (e.g. > 1 kV/mm). In contrast, the hard PZT-based ceramics, such as PZT-4S, are better employed in applications where the material is subjected to high electric fields and mechanical stress, where a lower probability of mechanical or electrical depolarisation is of importance. However, it should be noted that this move to a harder material comes at the expense of lower piezoelectric strains and d_{ij} coefficients. For sensor applications, ferroelectric ceramics such as PZT-5H are of interest, primarily due to the high piezoelectric d_{ij} and g_{ij} coefficients.

Table 3.14 provides a summary of the range of ferroelectric and piezoelectric materials and properties of interest. It is clear that ferroelectric materials such as PZT, PMN–PT and $BaTiO_3$ exhibit high piezoelectric d_{ij} coefficients compared to non-ferroelectric piezoelectric materials (such as GaN, ZnO, AlN, CdS and α-Quartz); as a result, ferroelectric materials are well suited to actuator applications. The lack of mobile ferroelectric domains in materials such as ZnO, AlN, CdS, GaN and α-Quartz leads to very large quality factors (Q_f), see Table 3.14, and therefore sharp resonances; this is why quartz is used in crystal resonators.

Note 3.10

Soft ferroelectrics have easy domain motion and, compared to hard ferroelectrics, have high d_{ij} coefficients, high losses, low coercive field (making them easier to pole), low depolarisation stress, low permittivity and low quality factor. They are attractive for sensing or low-field applications.

Note 3.11

Hard ferroelectrics have more restricted domain motion and, compared to soft ferroelectrics, have low d_{ij} coefficients, low losses, low permittivity, high coercive field (are more difficult to pole), higher depolarisation stress and higher quality factor. They are attractive for high-field, high-stress and/or resonant-based applications.

3.5.3 Ferroelectric Polymers

The majority of ferroelectric materials we have discussed are ceramics; however, ferroelectric polymers are also available, such as polyvinylidene fluoride (PVDF) and their co-polymers. However, they exhibit low stiffness, which often precludes them from high-force actuator applications. Nevertheless, their low permittivity leads to high g_{ij} coefficients (electric field generated per unit stress), making them suitable for sensors. As with the ferroelectric ceramics, their ferroelectric behaviour stems from the presence of a dipole, in this case, a molecular dipole, that can be switched by applying an electric field. To achieve a high level of polarisation, mechanical stretching is often used to align the polymer molecules. The reader is referred to a review on ferroelectric polymers [18].

3.6 Mechanical Depolarisation of Ferroelectric Materials

Since a piezoelectric transducer element can be subject to a range of mechanical stresses when it is used as either an actuator or a sensor, there is a need to understand the effect of an applied stress on piezoelectric and ferroelectric materials. The application of high levels of stress on piezoceramic materials can lead to *non-linear* stress–strain behaviour and a permanent plastic-like deformation of a ceramic material due to domain motion. The key factors which influence the response of a ferroelectric ceramic when it is subject to a high level of mechanical loading are:

i) the direction of the applied mechanical load, particularly in relation to the poling/polarisation direction, i.e. if the load is parallel (in the 3-direction) or perpendicular (in the 1- or 2-direction) to the polarisation axis.
ii) the properties of the piezoceramic material and ease of domain motion, i.e. whether the material is a ‘soft’-type or a ‘hard’-type ferroelectric material.

Table 3.14 Dielectric, piezoelectric, mechanical and pyroelectric properties of some common piezoelectric materials and nanowires (NW) Bowen et al. [17].

Material	GaN	AlN	CdS	ZnO	α-Quartz	$BaTiO_3$	PZT-4 'hard PZT'	PZT-5H 'soft PZT'	PMN–PT	$LiNbO_3$	PVDF
Piezoelectric	✓	✓	✓	✓	✓	✓	✓	✓	✓	✓	✓
Pyroelectric	✓	✓	✓	✓	✗	✓	✓	✓	✓	✓	✓
Ferroelectric	✗	✗	✗	✗	✗	✓	✓	✓	✓	✓	✓
Constant strain relative permittivity (ε_{33}^S)	11.2	10.0	9.53	8.84	4.63	910	635	1470	680	27.9	5–13
Constant stress relative permittivity (ε_{33}^T)	—	11.9	10.33	11.0	4.63	1200	1300	3400	8200	28.7	7.6
d_{33} pC/N	3.7 13.2 (NW)	5	10.3	12.4 14.3–26.7 (nanobelt)	$d_{11} = -2.3$	149	289	593	2820	6	−33
d_{13} pC/N	−1.9 −9.4 (NW)	−2	−5.18	−5.0	—	−58	−123	−274	−1330	−1.0	21
d_{15} pC/N	3.1	3.6	−13.98	−8.3	$d_{14} = 0.67$	242	495	741	146	69	−27
Mechanical quality (Q_m)	2800 (NW)	2490	~1000 (NW)	1770	10^5–10^6	400	500	65	43–2050	10^4	3–10
Electromechanical coupling (k_{33})	—	0.23	0.26	0.48	0.1	0.49	0.7	0.75	0.94	0.23	0.19
$-p$ (μC/m²/K)	4.8	6–8	4	9.4	—	200	260, 533 (PZT)	260–533 (PZT)	1790	83	33
S_{11}^E (p/Pa)	3.326	2.854	20.69	7.86	12.77	8.6	12.3	16.4	69.0	5.83	365
S_{33}^E (p/Pa)	2.915	2.824	16.97	6.94	9.73	9.1	15.5	20.8	119.6	5.02	472

Source: Adapted from Kim et al. (2014).

Ferroelectric-based piezoelectric ceramics that have been poled and have a remnant polarisation are observed to exhibit a degree of non-linear deformation when they are subjected to high compressive loads. In this regard, a compressive load applied in a direction that is parallel to the poling direction can lead to a loss of polarisation (depolarisation) and loss of piezoelectric properties, while a load applied in a direction perpendicular to the poling direction can stabilise the degree of polarisation. The non-linear deformation of a ferroelectric material, such as a ceramic, has been related to the depolarisation and stress-induced domain switching of the material; this can be considered to be a ferro-elastic response, rather than a ferroelectric response [3, 19–21].

Cao and Evans [19] have examined in detail the stress-induced depolarisation of PZT-based ferroelectric materials; this was undertaken with simultaneous measurements of the stress–strain behaviour and measurements of the piezoelectric voltage that was generated across a capacitor connected in parallel with the material being loaded. The electrical measurements enabled the polarisation charge to be quantified during loading, and this was correlated to the mechanical stress applied to the ferroelectric material; see Figure 3.9. The testing cycle involved loading and unloading ferroelectric ceramics in compression within a relatively small range of stresses (<5 MPa), which gradually increased up to a maximum stress of over 500 MPa. The application of compressive load is typical for ceramic materials and is easier to undertake compared to tensile testing. The hard and soft PZTs were examined since the materials exhibit different levels of domain mobility (hard–difficult, soft–easy domain motion).

The testing identified that when the compression stress was greater than 300 MPa, mechanical failure of PZT was observed, and the critical stress levels where non-linearities are observed in the stress–strain behaviour are different between the hard PZT and soft PZT materials. When the compressive load was applied in a direction that was parallel to the poling direction, the soft PZT material exhibited non-linear stress–strain behaviour (due to domain switching) when the stress was greater than ~20 MPa. In contrast, the hard PZT exhibited a far more linear stress–strain response at higher levels of stress, up to 60 MPa. This observation is in good agreement with the assumption that non-linearities in the stress–strain behaviour are correlated with domain motion since the hard PZT has more restricted domain motion (see Figure 3.8a). When the stress was removed, some degree of plastic/permanent strain was observed due to domain motion; this unrecoverable deformation was observed for the soft PZT material at relatively low-stress levels of ~30 MPa. However, the hard PZT material was able to recover its elastic deformation at stress levels up to 300 MPa since domain motion was more limited in this stress range. The application of a very high compressive stress, in excess of 300 MPa, resulted in the hard PZT permanently deforming, leading to permanent strain of the material (Figure 3.8a). The critical stress level at which ferroelectric domains begin to switch and move can be termed *coercive stress*, which is analogous to the coercive field.

The impact of a compressive stress on a poled ferroelectric along the poling direction can be considered by imagining the application of a compressive stress along the polarised 'up' or 'down' crystal structure in Figure 2.2a. The compressive stress will reduce the length of the unit cell in the poling direction so that it becomes more cubic and the central Ti^{4+} or Zr^{4+} moves towards the centre so that eventually switching of the dipole and domains can occur due to the applied mechanical stress.

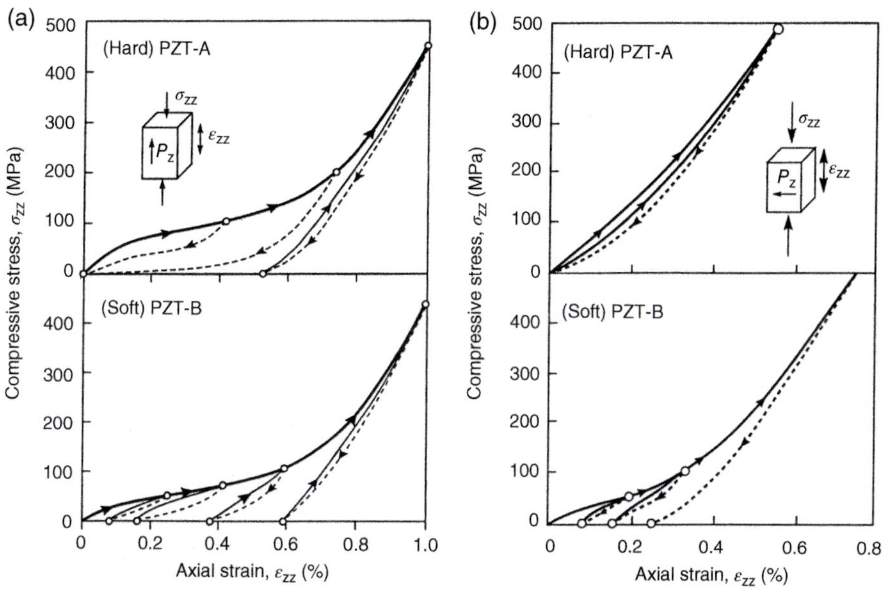

Figure 3.8 Stress–strain curves for both a hard PZT and a soft PZT that has been subjected to (a) a mechanical stress that is along the polar axis and (b) a mechanical stress that is applied normal to the polar axis. *Source:* Cao and Evans [19] / with permission of John Wiley & Sons.

The application of compressive loads that are perpendicular to the poling direction has a different effect; perpendicular loading leads to limited levels of non-linear stress–strain behaviour; see Figure 3.8b. The application of small levels of strain (or stress) in a direction perpendicular to the polarisation can increase the level of polarisation of the PZT, but a further increase in strain leads to the eventual depolarisation of ferroelectric. As a result, the application of a compressive stress that is perpendicular to the polarisation direction has a limited effect on the domain alignment. The use of doping of the ferroelectric ceramic and degree of ferroelectric 'hardness' (e.g. selection of a hard or soft PZT) in terms of domain motion is therefore seen to strongly effect the behaviour under mechanical loading, and it is no surprise that hard PZT materials are used in applications where high levels of stress are applied (particularly along the poling axis).

The impact of a compressive stress on a poled ferroelectric that is applied in a direction that is perpendicular to the poling direction can also be considered by imagining the application of a compressive stress perpendicular to the polarised 'up' or 'down' crystal structure in Figure 2.2a. In this case, the perpendicular compressive stress will increase the length of the unit cell in the poling direction, and so, a limited level of depolarisation or permanent deformation is observed in Figure 3.8b.

Note 3.12

High levels of stress can lead to depolarisation of the ferroelectric material, and the ease of depolarisation depends on the direction of stress relative to the polarisation direction and the type of material (hard or soft).

3.7 Creep of Ferroelectric Materials

Piezoelectric actuators [3, 22–24] can exhibit a form of 'creep' in their deflection, where there is a time-dependent displacement of a piezoelectric material when it is subject to a DC voltage excitation. Here, we will outline the fundamental mechanism that results in this time-dependent creep-like behaviour of a piezoelectric material. The instantaneous strain that is produced by a piezoelectric is understood by the constitutive equations, including those described earlier in the introductory section of this chapter. However, the time-dependent displacement creep of a piezoelectric actuator, which is observed under open-loop voltage control [22], is a result of *extrinsic* effects that are linked to time-dependent domain reorientation. Figure 3.9 is a diagram of the unit cell of a PZT ceramic and the resulting strain in the polarisation direction, which is produced in response to the applied electric field [26]. The initial rapid extension of the dipole is an example of an *intrinsic* response to produce a piezoelectric strain.

The *extrinsic* (and time-dependent) response to an electric field can be considered at a scale larger than a single unit cell of a ferroelectric ceramic. The two main contributions that lead to the time-dependent creep of a ferroelectric/piezoelectric material are (i) domain wall motion and (ii) local phase transitions. It is these two extrinsic effects that lead to the gradual creep and deflection of an actuator when it is subject to simple open-loop control (e.g. the application of a constant voltage to an actuator for a prolonged period). We have seen in Chapters 1 and 2 that ferroelectric domains are regions in the material that share a common polarisation axis. Domain wall motion can therefore be characterised by a boundary between adjacent ferroelectric domains that move in response to the applied electric field, which results in a reduction in the volume of domains that have a polarisation axis which is poorly aligned with the applied electric field. Adjacent ferroelectric domains, whose polarisation axis is better aligned with the applied field, will increase in volume, as shown in Figure 3.10, where 90° domain motion leads to an increased level of polarisation.

As the polarisation axis gradually becomes more aligned with the direction of the applied electric field with time, the induced piezoelectric strain of the ferroelectric will increase since reoriented domains will exhibit a greater piezoelectric strain compared to less well-

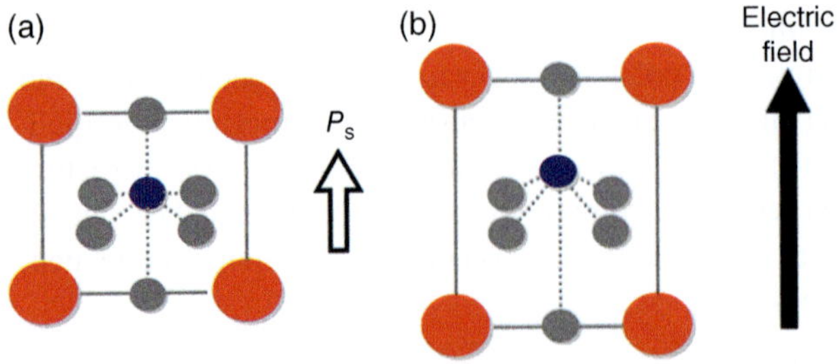

Figure 3.9 Intrinsic strain that is induced in a tetragonal PZT unit cell with a polarisation axis (P_s) before (a) and after the application of an applied electric field (b). *Source:* Giddings [25] / with permission of Peter Francis Giddings.

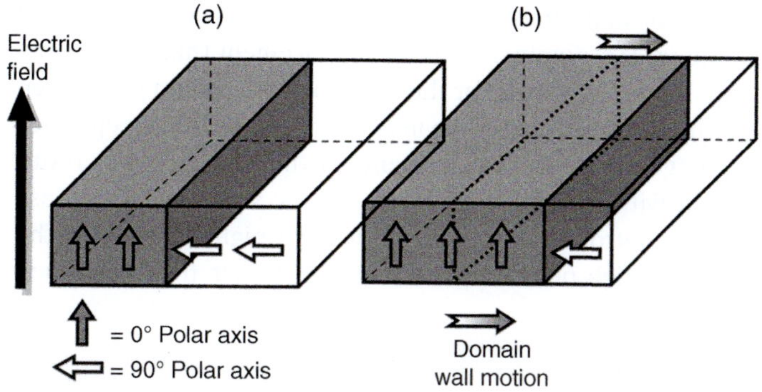

Figure 3.10 Polarisation of ferroelectric domains immediately after the application of an electric field (a) and at a later time *t* (b), showing domain wall motion, thereby increasing the size of well-aligned domains. *Source:* Giddings [25] / with permission of Peter Francis Giddings.

orientated domain; this is somewhat similar to the gradual increase in strain as the electric field is increased, as described by the strain–field response in Figure 3.7. This time-dependent increase in domain alignment, and therefore induced piezoelectric strain, is observed as creep displacement of a piezoelectric when it is subject to a constant applied field or voltage [22–24].

If the ferroelectric has a tetragonal crystal structure, the nature of the domain wall motion shown in Figure 3.10 is primarily a consequence of the reorientation of 90° domains [27]. In addition to domain wall motion, domains with a polar axis that are still highly misaligned, relative to the applied field, may also be reoriented by a change in phase of the material within domains; this is achieved by local phase changes of the material. As a result of a change in phase, such as from a tetragonal crystal structure to a rhombohedral structure, domains can be better aligned due to a change in the polar axis of 109° or 71° [9] to achieve improved alignment of the polar axis with the corresponding driving field. A localised phase transition can be achieved in PZT compositions that are close to the morphotropic phase boundary, where there are almost equal proportions of a titanium-rich rhombohedral phase and a zirconium-rich tetragonal phase. In these ceramic compositions, the application of an electric field provides sufficient energy to initiate a phase transition. Based on these discussions, it would be expected that a soft PZT type, which exhibits easier domain motion, is likely to exhibit greater levels of creep compared to the hard PZT materials. Similarly, non-ferroelectric piezoelectric materials, for example quartz or GaN, are expected to exhibit less, or no, creep since there are no ferroelectric domains.

A number of creep models have been proposed to quantify the time-dependent creep displacement of a piezoelectric material; in this regard, the model provided by Jung [24], Eq. (3.25), has been used to predict the creep response in PZT materials.

$$x(t) = x_0\left[1 + \gamma \log\left(\frac{t}{0.1}\right)\right] \tag{3.25}$$

where $x(t)$ is the displacement (or deflection) of the piezoelectric actuator for a fixed applied voltage or electric field, x_0 is a nominal constant displacement (namely the displacement 0.1 s after applying the input voltage for the equation above), and γ is a creep factor that describes the logarithmic creep rate of the material or device (the creep factor can be considered as a percentage change in displacement per time decade). As a result of this creep response, coupled with the hysteretic nature of the strain–field response seen in Figure 3.7, closed-loop control is often used to achieve high accuracy in the use of piezoelectric actuators for high-precision displacement, rather than open-loop control.

Note 3.13

Under open-loop conditions, a ferroelectric actuator can be subject to creep as a result of time-dependent domain motion. Soft ferroelectrics are also likely to exhibit greater levels of creep, compared to hard ferroelectrics, due to their easier domain motion.

3.8 Strain Limits of Piezoelectric Actuators (Expansion)

The maximum strain reported for piezoelectric actuators is generally the expansion strain (often termed 'free stroke' or 'free displacement' with no applied load). This corresponds to the application of a maximum electric field (and voltage) that is parallel to the poling direction of the material in order to extend the dipole (and even lead to some domain motion). The level of electric field is governed by the need to minimise long-term damage/fatigue of the materials and/or dielectric breakdown, and typical maximum electric fields can be 1–3 kV/mm. This can be considered as operating in the region D → C in Figure 3.7.

There is a limit to the degree of strain that a piezoelectric material, such as PZT, can generate. The amount of strain depends on the material, but most polycrystalline PZT piezoelectric materials range from 0.1% to 0.3% [26]. When the material is incorporated into an actuator device, there may be a reduction in the maximum strain due to the electrode structure and distribution of the applied electric field; for example, in a multilayer structure, the maximum strain can be ~0.1–0.15%. A simple rule of thumb for 0.1% strain is a piezoelectric displacement of 1 μm for every 1 mm length of an actuator material.

Single-crystal ferroelectric materials are available that exhibit larger maximum strains, for example relaxor ferroelectric solid-solution single crystals which include $Pb(Mg_{1/3}Nb_{2/3})O_3$–$PbTiO_3$ (PMN–PT) and $Pb(Zn_{1/3}Nb_{2/3})O_3$–$PbTiO_3$ (PZN–PT). Such ferroelectric crystals exhibit piezoelectric d_{33} coefficients of 1200–2500 pC/N; this is significantly larger than typical polycrystalline ferroelectric/piezoelectric ceramics such as the soft PZT ceramics whose d_{33} coefficient is 500–700 pC/N. New Sm-doped $Pb(Mg_{1/3}Nb_{2/3})O_3$–$PbTiO_3$ (Sm-PMN–PT) single crystals have recently been reported with even further improvements in performance [28].

Note 3.14

Maximum elongation strain is typically ~0.3% for PZT-ceramic-based materials and 0.1–0.15% for piezoelectric actuators. For example, a piezoelectric multilayer stack of height ~ 70–100 mm is required for free deflection of ~100 μm. A simple rule of thumb for ~0.1% strain is a piezoelectric displacement of 1 μm for each 1 mm length of an actuator material. Single-crystal ferroelectric materials are now available, which can exhibit higher strains and *d*-coefficients.

3.9 Strain Limits of Piezoelectric Actuators (Contraction)

As stated in Chapter 2, it is also possible to apply an electric field in a direction that is opposite to the poling direction in an effort to contract the piezoelectric actuator; however, there is a limit to the degree of contraction which can be attained and it is less than the amount of extension. If a high electric field is applied in a direction that is opposite to the poling direction, it is possible to lose the polarisation (depolarisation) or even reverse the poling direction (domain switching); reversal of the poling direction would lead to a subsequent *expansion* of the ceramic since the polar axis is now in the electric field direction. This effect can be observed by consideration of the strain–field loop in Figure 3.7, where operating in the region D → E leads to contraction; however, if the magnitude of the electric field exceeds the coercive field of the material (E_c) at point E in Figure 3.7, operating in the region E → F will lead to expansion.

The highest strain ferroelectric materials, which are chosen for piezoelectric actuators, are often 'soft' compositions, where domain motion and switching are relatively easy. As a result, the degree of contraction is often limited to ~20–30% of the free stroke expansion; control of the hysteresis to achieve high-precision actuation is achieved using some form of closed-loop control.

Note 3.15

The degree of contraction is limited to 20–30% of expansion to prevent loss or switching of the polarisation. For example, when the free stroke expansion is 100 μm, the contraction will be limited to 20–30μm.

3.10 Resonance Behaviour of Piezoelectric Materials and Ceramic Structures

We have seen that piezoelectric materials can be employed for operation at a low frequency (approaching 0 Hz); however, they can also operate at high frequencies and at the resonant frequency of the device/structure [3]. Piezoelectric ceramics can be manufactured in a wide variety of shapes, while crystals will need to be cut into specific geometries (often termed 'cuts') to resonate with specific modes and at specific frequencies. Resonant applications of piezoelectric include sensors (e.g. quartz crystal microbalances to measure mass), actuators/transmitters (e.g. ultrasonic baths) and non-destructive testing [10, 29] (e.g. transmit and receive transducers for pulse-echo), where the piezoelectric material can be excited using an AC electric field to produce a vibrational response at a specific resonant frequency. The mode of resonance and the resonant frequency of a piezoelectric element are related to its geometry (dimensions) and material properties (density and stiffness). Applications of piezoelectric elements operating at resonance, such as cylinders, can include sonar transmitters, micro-positioning that uses inch-worm-type actuation at resonance and scanning microscopy systems.

The response of a piezoelectric element as a result of excitation at a resonance frequency can be characterised by the modal shapes that are produced during excitation; see Figure 3.11a. The nature of the harmonic response can be experimentally investigated using

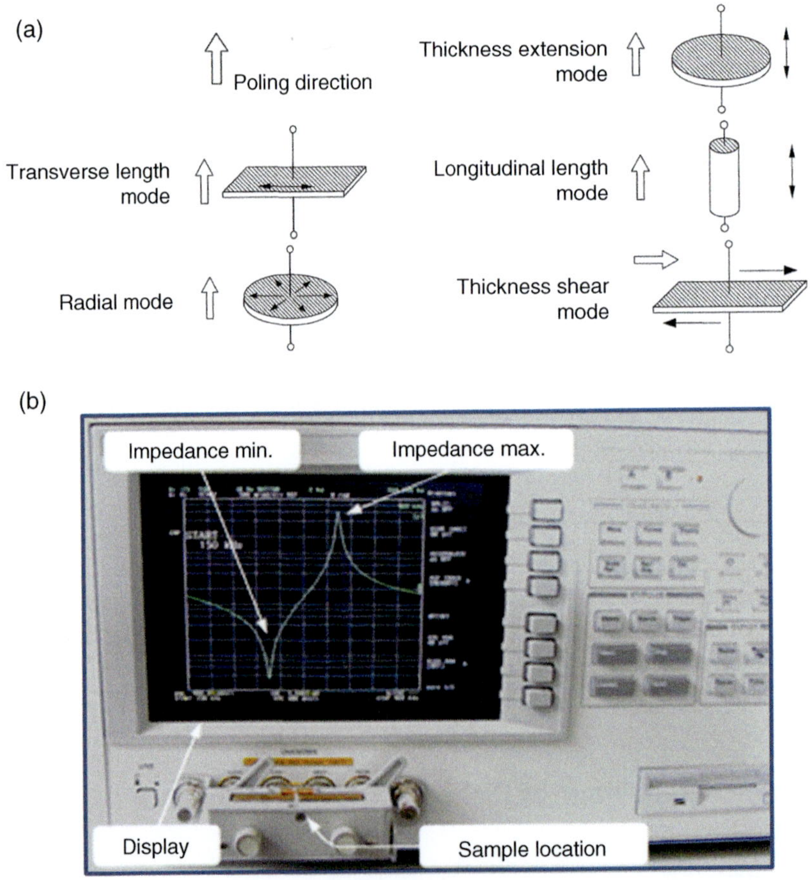

Figure 3.11 (a) Fundamental vibration modes for a piezoelectric material; the grey areas indicate electrode positions, and arrows indicate poling direction. (b) Example of experimental configuration form impedance response of a piezoelectric ceramic, with an impedance-versus-frequency-generated curve. *Source:* With permission of Rhodri Lewis.

impedance analysis, which measures the electrical response of the piezoelectric element (e.g. the real and imaginary impedance) across a frequency range [30, 31]; see Figure 3.11b. Such data can be used to determine the quality factor (Q_f), which is a measure of the sharpness of the resonance.

The forced vibration and resonance of a piezoelectric material that is excited by an AC electric field near its natural frequency can be defined as a *resonance frequency* and can be defined for a simple structure by Eq. (3.26):

$$f_r = \frac{C}{2\pi L} \tag{3.26}$$

where f_r is defined as the resonance frequency of the piezoelectric element, L is the length of the piezoelectric element, and C is the speed of sound of the piezoelectric material (which will depend on its density and stiffness). An increase in the stiffness of the material will lead to an increase in the speed of sound, which will therefore increase the resonant frequency. Likewise, if the length of the material/device is reduced, it will lead to an increase in the resonant frequency since the first mode of resonance occurs when the length is equal to an integral of a half-wavelength [32].

An excellent report on the measurement of piezoelectric resonance has been produced by the National Physical Laboratory [33].

3.11 Ageing of Ferroelectrics

We have seen that ferroelectric materials are poled to align domains in a common direction and achieve a remnant polarisation. However, a random domain orientation corresponds to a lower energy state due to a reduced net electrical potential and depolarising field across a crystalline solid; see Figure 2.5. Therefore, poled piezoelectric materials often undergo some form of ageing; this can be described as a spontaneous change in an electromechanical property with time when subject to zero external stress and constant temperature. It also corresponds to a loss of the remnant polarisation, shown in Figure 3.5, with time [34]. The ageing rate (A) is logarithmic since it is a thermally activated process and is often described as *% change per time decade* at a particular temperature.

$$A = \frac{1}{(\log(t_2) - \log(t_1))}\left(\frac{P(t_2) - P(t_1)}{P(t_1)}\right) \tag{3.27}$$

where P is a property of interest, such as d_{ij}, permittivity or coupling coefficient, and $P(t_1)$ is the magnitude at time t_1 (e.g. a d-coefficient measured at time t_1 after poling), $P(t_2)$ is the magnitude at any subsequent time t_2, and A is a parameter termed the *ageing per time decade* [35]. As an example, the ageing of the piezoelectric d-coefficient can be approximately −3% per time decade for PZT at room temperature, with a negative value indicating a decrease in d-coefficient with time. Thus, from day 1 to 10, the d-coefficient will decrease by 3%; it will decrease another 3% by day 100, and then, it will decrease another 3% by day 1000, etc.

Many piezoelectric and dielectric properties are measured at least 24 hours after poling since the largest change occurs directly after poling; for example, if the piezoelectric *d*-coefficient of a PZT with an ageing rate of approximately −3% per time decade was measured at 1 minute after poling, it would decrease by 3% after 10 minutes, another 3% by 100 minutes and then another 3% after 1000 minutes (16.6 hours).

An excellent review of mechanisms in ageing and fatigue is available for more detailed information [36].

3.12 Temperature Limits and Self-Heating

As stated in Chapter 2, if a ferroelectric ceramic is heated to a temperature above its Curie point (T_c), it transforms to the cubic phase. It is therefore no longer ferroelectric or piezoelectric, and there are no domains since above this temperature the crystal structure is symmetrical. On cooling back down below the Curie point of the material, the domains will reappear; however, they will form in a random configuration over the volume of the material to reduce the net electric potential and minimise strain. This will result in a complete loss of actuator or sensor performance. The Curie point therefore represents an absolute maximum temperature of operation. Manufacturers often recommend an upper operating temperature that is approximately half the Curie point (in units of °C). For example, piezoelectric actuators with a Curie point of 300 °C can be used at a temperature up to 150 °C and actuators with a Curie point of 160 °C can be operated at up to 80 °C. For operation at high temperatures, for example in oil, gas, engine or industrial applications, a variety of high T_c materials are available [37].

A lower operating temperature is less of a problem, and piezoelectric materials have been used down to cryogenic temperatures. At very low temperatures (e.g. greater than −170 °C), there is some decrease in the amount of strain due to the lack of domain mobility at such low temperatures; data can be found in this excellent report [38].

Note 3.16

The Curie point (T_c) represents an absolute maximum temperature of operation since above this temperature the material is no longer ferroelectric. The recommended maximum operating temperature of a piezoelectric material is typically half the Curie temperature (in units of °C). For example, the Curie point of the ferroelectric should be ~300–350 °C for operation at 150 °C.

3.13 Cyclic Operation – Frequency Effects

The frequency of operation of a piezoelectric transducer can be high for piezoelectric materials with actuators operating in the Hz-to-kHz regime and piezoelectric ultrasonic transducers operating in the MHz range. Two main issues with regard to the operation of piezoelectric actuators under AC conditions are firstly the possible heating of the actuator due to hysteresis, which can lead to thermal runaway and subsequent loss of polarisation, and secondly the electric current and voltage requirements to drive the actuator. These are now discussed.

3.13.1 Self-Heating Due to Ferroelectric Hysteresis

We have seen that ferroelectric materials exhibit some form of hysteresis as a result of domain motion. In Figure 3.5, complete domain reversal was described, but a finite level of hysteresis can be achieved even under unipolar operation or at electric fields below the coercive field. Hysteresis losses can result in self-heating of the material, especially when operating with high electric fields at high frequencies, and self-heating can facilitate even greater domain motion leading to thermal runaway. Therefore, it is of interest to consider heating during operation.

Zheng et al. [39] outlined an approach to examine the generation of heat in piezoelectric ceramic actuators [39], where the temperature increase of a piezoelectric actuator can be estimated from:

$$\Delta T = \frac{u \cdot f \cdot v_e}{k(T) \cdot A} \tag{3.28}$$

where ΔT is the temperature increase (*K*); *u* is the total dielectric loss of the piezoelectric material per unit volume per driving cycle (J/mm^3); this is primarily a result of ferroelectric hysteresis (as shown in Figure 3.5); *f* is the driving frequency (Hz); v_e is the active volume of the actuator (m^3); *A* is the total surface area of the actuator (m^2); and $k(T)$ is the overall heat transfer coefficient of the system (W/m^2/K) [4].

While Eq. (3.28) outlines the increase in steady-state temperature of the piezoelectric actuator, the time–temperature profile of the piezoelectric actuator device can be described by [4]:

$$T = T_0 + \Delta T\left(1 - e^{-t/\tau}\right) \tag{3.29}$$

where *t* represents the instantaneous time (*s*), τ is a time constant (s), T_0 is the initial temperature (*K*), *T* is the actuator temperature at a time *t* (*K*), and ΔT is the maximum temperature increase of the piezoelectric actuator (*K*). From Eq. (3.29), at the initial condition, $t = 0$; then, the temperature $T = T_0$; then, at a condition of equilibrium when the time, $t \rightarrow \infty$; and then, the temperature increases to $\mathrm{T} = \mathrm{T_0} + \Delta\mathrm{T}$. The time constant ($\tau$), shown by Eq. (3.29), can be described by

$$\tau = \frac{v \cdot \rho \cdot c}{k(T) \cdot A} \tag{3.30}$$

where v is the total volume of the actuator (m^3), ρ is the actuator density (kg/m^3), and *c* is the specific heat capacity of the actuator material (J/kg/K). An example system will be described later in the book as a case study in Chapter 8.

It is often difficult to know the heat transfer coefficient, but the worst case is when the actuator is suspended in air and Zheng et al. reported a $k(T)$ of 20–40 W/m^2/K for such a condition. Figure 3.12 shows the temperature increase versus frequency for a small multilayer actuator attached to steel with a high heat transfer coefficient [4]. As can be seen, for frequencies <50 Hz, the degree of heating is small (< 2 °C) and substantial heating is only

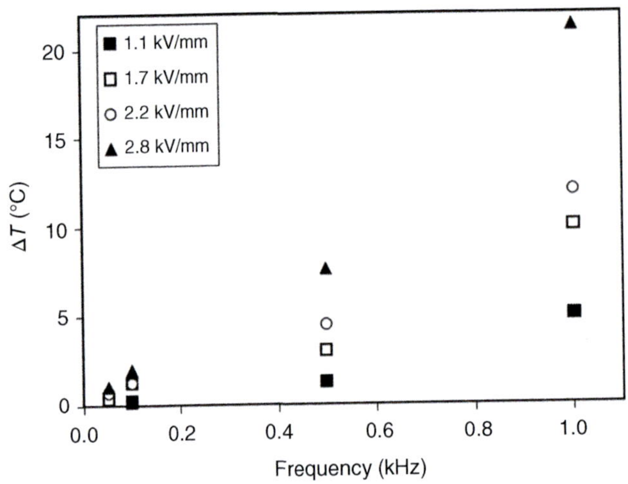

Figure 3.12 Temperature increase versus frequency for multilayer actuator. *Source:* Pritchard et al. [4] / with permission of Elsevier.

attained at frequencies of 500 Hz and above. Once an actuator is selected, Eq. (3.25) can be used to provide some estimate of temperature increase based on u of $\sim$2 kJ/m^3 [4] and $k(T)$ of 30 W/m^2/K [4]. Clearly selecting materials with lower hysteresis losses will lead to reduced levels of heating, such as the more 'hard' ferroelectrics in Table 3.12.

3.13.2 Current Requirements During Frequency Cycling

Piezoelectric materials are good insulators (they are *dielectrics*), and when an electric field is applied under DC conditions, the electric current can be considered to be relatively small. However, this does not imply that the necessary current to drive an actuator is always small since under AC conditions the current requirement depends on (i) the frequency of operation, (ii) the driving voltage and (iii) the capacitance of the actuator. Cycling a piezoelectric actuator can be regarded as charging and discharging a capacitor, a *capacitive* load. The high degree of polarisation of ferroelectric materials results in materials of high relative permittivity (in excess of 1000). Therefore, the capacitance of piezoelectric actuators can be large and the necessary current can also be large, especially if the actuator is being operated at a high frequency and voltage. As a result, the drive electronics for piezoelectric actuators are generally aimed at driving capacitive loads.

As an example, we will consider a Noliac SCMA-R 5 actuator, which is 200 mm in height and has a free stroke of 282.2 μm at 200 V. The actuator has a capacitance of 77 600 nF. Based on these details, an estimate of the electric current to drive the actuator sinusoidally can be determined from Eq. (3.31).

$$i_{maximum} \sim \pi \cdot f \cdot C \cdot V \tag{3.31}$$

where f is the operating frequency, C is the piezoelectric device capacitance, V is the maximum voltage, and $i_{maximum}$ is the peak current. Using the data above, a current amplitude of $\sim$1 A is required to drive the actuator at 18 Hz, demonstrating that with AC, high currents can be required.

Since the actuators are capacitors which are charged and discharged as they are expanded and contracted, it may also be possible to use charge recovery to reduce power consumption [40]. As an example of a commercial system, some piezoelectric amplifiers employ pulse width modulation, along with some form of recovery of the electrical energy. As an example, instead of simply employing a heat sink to dissipate reactive power, electrical energy is stored within inductive elements to power the piezoelectric actuator; as a result, any energy that is subsequently not used by the piezoelectric actuator can then be returned to the amplifier and used again.

Note 3.17
Piezoelectric transducers can be considered as a capacitive electric load. The necessary current increases with frequency, drive voltage and device capacitance.

3.14 Thermal Expansion Coefficient

It may be necessary to know the thermal expansion coefficient of the ferroelectric material used, for example, to estimate thermal strains or any stress in a system due to thermal expansion mismatches. Since the material is polarised, the thermal expansion coefficient of a poled ferroelectric material can be anisotropic and can range from negative values $-4 \times 10^{-4}\%/°C$ to positive values $8 \times 10^{-4}\%/°C$ and this depends on the material. This complex behaviour of poled ferroelectric ceramics originates from a coupling between the temperature-dependent remnant polarisation of the material and the strain. A positive expansion is associated with conventional thermal expansion, and a negative expansion is associated with the change from a highly tetragonal (elongated) crystal to a cubic structure, which is stable at the higher temperature ($>T_c$), i.e. the volume decrease during the ferroelectric-to-paraelectric phase transition. Figure 3.13 shows thermal strain versus

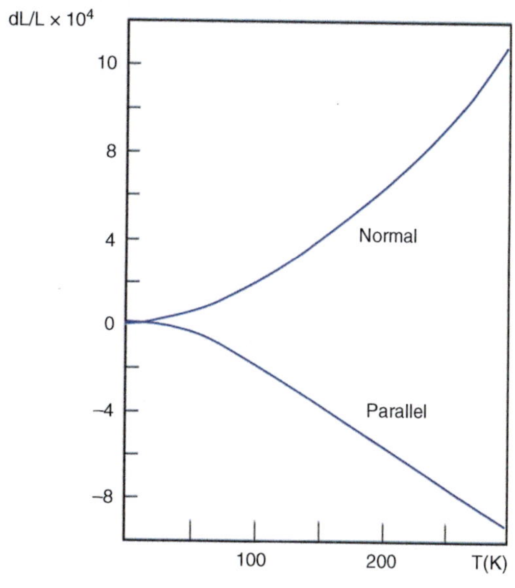

Figure 3.13 Thermal strain versus temperature in a lead zirconate titanate material. *Source:* Simpson and Wolfs [41] / with permission of AIP Publishing LLC.

temperature, showing a negative thermal strain parallel to the polarisation direction and a positive thermal strain normal to the polarisation direction.

An excellent overview of the most important methods employed in the characterisation of piezoelectric, ferroelectric and pyroelectric materials has been edited by Cain [42]; this includes a detailed overview of measurement best practices, worked examples and in-depth descriptions of fundamentals and methods.

3.15 Summary

The fundamental behaviour of piezoelectric and ferroelectric materials has been described; this information provides important guidance to understanding the operational limits and selection of the best materials for specific piezoelectric applications. The following chapters will now describe their application in a variety of actuator (Chapter 4), sensor (Chapter 5) and energy harvesting (Chapter 6) applications.

References

1 Hall, D.A. (2001). Review nonlinearity in piezoelectric ceramics. *Journal of Materials Science* 36: 4575–4601.

2 Dent, A. C. (2008). Novel active fibre composites. PhD thesis. University of Bath.

3 Lewis, R. (2012). Innovation of a Piezoelectric Touch Trigger Probe. PhD thesis. University of Bath.

4 Pritchard, J., Ramesh, R., and Bowen, C.R. (2004). Time–temperature profiles of multi-layer actuators. *Sensors and Actuators A* 115: 140–145.

5 Berlincourt, D. (1981). Piezoelectric ceramics - Characteristics and applications. *Journal of the Acoustical Society of America* 70: 1586–1595.

6 Steinhausen, R., Pientschke, C., Kuvatov, A.Z. et al. Modelling and characterization of piezoelectric and polarization gradients. *Journal of Electroceramics* 20: 47–52.

7 Jin, L., Fei, L., and Shujun, Z. (2014). Decoding the fingerprint of ferroelectric loops: Comprehension of the material properties and structures. *Journal of the American Ceramic Society* 97: 1–27.

8 Scott, J.F. (2008). Ferroelectrics go bananas. *Journal of Physics: Condensed Matter* 20: 021001.

9 Haertling, G.H. (1999). Ferroelectric ceramics: history and technology. *Journal of the American Ceramic Society* 82: 797–818.

10 Tressler, J.F., Alkoy, S., and Newnham, R.E. (1998). Piezoelectric sensors and sensor materials. *Journal of Electroceramics* 2: 257–272.

11 Berlincourt, D. (1992). Piezoelectric ceramic compositional development. *Journal of the Acoustical Society of America* 91: 3034–3040.

12 Cowley, R.A., Gvasaliya, S.N., Lushnikov, S.G. et al. (2011). Relaxing with relaxors: a review of relaxor ferroelectric. *Advances in Physics* 60: 229–327.

13 Gerson, R. and Jaffe, H. (1963). Electrical conductivity in lead titanate zirconate ceramics. *Journal of Physics and Chemistry of Solids* 24: 979–984.

14 DOD-STD-1376A. (1984). Military Standard: Piezoelectric Ceramic Material and Measurements Guidelines for SONAR Transducers.

15 Berlincourt, D. A, Krueger, H. H. A., Near, C, Properties of Piezoelectricity Ceramics. Morgan Electro Ceramics Technical Publications - TP226.

16 Bent, A (1997). Active fiber composites for structural actuation. PhD Thesis. Massachusetts Institute of Technology.

17 Bowen, C.R., Kim, H.A., Weaver, P.M., and Dunn, S. (2014). Piezoelectric and ferroelectric materials and structures for energy harvesting applications. *Energy & Environmental Science* 7 (1): 25–44.

18 Chaoying, W. and Bowen, C.R. (2017). Multiscale-structuring of polyvinylidene fluoride for energy harvesting: the impact of molecular-, micro- and macro-structure. *Journal of Materials Chemistry A* 5: 3091.

19 Cao, H.C. and Evans, A.G. (1993). Nonlinear deformation of ferroelectric ceramics. *Journal of the American Ceramic Society* 76: 890–896.

20 Krueger, H.H.A. (1967). Stress sensitivity of piezoelectric ceramics: part 1. Sensitivity to compressive stress parallel to the polar axis. *Journal of the Acoustical Society of America* 42: 636.

21 Krueger, H.H.A. (1968). Stress sensitivity of piezoelectric ceramics: part 3. Sensitivity to compressive stress perpendicular to the polar axis. *Journal of the Acoustical Society of America* 43: 583.

22 Jung, H. and Gweon, D.G. (2000). Creep characteristics of piezoelectric actuators. *Review of Scientific Instruments* 71: 1896–1900.

23 Changhai, R. and Lining, S. (2005). Hysteresis and creep compensation for piezoelectric actuator in open-loop operation. *Sensors and Actuators A: Physical* 122: 124–130.

24 Jung, H., Shim, J.Y., and Gweon, D.G. (2001). Tracking control of piezoelectric actuators. *Nanotechnology* 12 (1): 14–20.

25 Giddings, P.F. (2010) Piezoelectrically Actuated Bistable Composite Laminates for Structural Morphing. PhD thesis. University of Bath.

26 Uchino, K. (2014). Piezoelectric actuator renaissance. *Energy Harvesting and Systems* 1: 45–56.

27 Huang, H. and Tianshu, Z. (2002). Stress and strain-relaxation in lead zirconate titanate based ceramics. *Materials Chemistry and Physics* 75: 186–189.

28 Li, F., Cabral, M.J., Xu, B. et al. (2019). Giant piezoelectricity of Sm-doped $Pb(Mg_{1/3}Nb_{2/3})O_3$-$PbTiO_3$ single crystals. *Science* 364: 264–268.

29 Tressler, J.F., Alkoy, S., Dogan, A., and Newnham, R.E. (1999). Functional composites for sensors, actuators and transducers. *Composites Part A -Applied Science and Manufacturing* 30: 477–482.

30 Huang, C.H., Lin, Y.C., and Ma, C.C. (2004). Theoretical analysis and experimental measurement for resonant vibration of piezoceramic circular plates. *IEEE Transactions on Ultrasonics Ferroelectrics and Frequency Control* 51: 12–24.

31 Kunkel, H.A., Locke, S., Pikeroen, B. et al. (1990). Finite-element analysis of vibrational modes in piezoelectric ceramic disks. *IEEE Transactions on Ultrasonics, Ferroelectrics, and Frequency Control* 37: 316–328.

32 Bedford, A. and Drumheller, D.S. (1994). *Introduction to elastic wave propagation.* London: Wiley.

33 Cain, M. and Stewart, M. (2001). *Piezoelectric Resonance*. Crown Copyright: NPL https://www.npl.co.uk/special-pages/guides/gpg33_resonance.

34 Gotmare, S.W., Leontsev, S.O., and Eitel, R.E. (2010). Thermal degradation and aging of high-temperature piezoelectric ceramics. *Journal of the American Ceramic Society* 93: 1965–1969.

35 Xu, Y. (1991). *Ferroelectric materials and their applications*. Elsevier.

36 Genenko, Y.A., Julia Glaum, J., Hoffmann, M.J., and Albe, K. (2015). Mechanisms of aging and fatigue in ferroelectrics. *Materials Science and Engineering B* 192: 52–82.

37 Damjanovic, D. (1998). Materials for high temperature piezoelectric transducers. *Current Opinion in Solid State and Materials Science* 3: 469–473.

38 Hooker, M.W. (1998). *Properties of PZT-based piezoelectric ceramics between −150 and 250 °C, NASA/CR-1998-208708*. Hampton, Virginia: Langley Research Center.

39 Zheng, J., Takahashi, S., Yoshikawa, S. et al. (1996). Heat generation in multilayer piezoelectric actuators. *Journal of the American Ceramic Society* 79: 3193.

40 Main, J.A., Newton, D.V., Massengill, L., and Garcia, E. (1996). Efficient power amplifiers for piezoelectric applications. *Smart Materials and Structures* 5: 766–775.

41 Simpson, A.M. and Wolfs, W. (1987). Thermal expansion and piezoelectric response of PZT channel 5800 for use in low temperature scanning tunneling microscope designs. *Review of Scientific Instruments* 58: 2193.

42 Cain, M.G. (ed.) (2014). *Characterisation of Ferroelectric Bulk Materials and Thin Films*, Springer Series in Measurement Science and Technology (SSMST, volume 2). Dordrecht: Springer.

4

Piezoelectric Actuators

4.1 Introduction

Piezoelectric actuators convert electrical energy into mechanical work and thus are able to exert forces and output motion. They have the potential for a high dynamic response, changing force and displacement at high frequency, within the limitations of the drive amplifier and the mechanical load characteristics. Most piezoelectric actuators can be classified as:

a) Axial actuators, using d_{33}-mode (e.g. stack actuators)
b) Flexural actuators, using d_{13}-mode (e.g. rectangular, ring and disc benders)

4.2 Free Displacement and Blocking Force

A piezoelectric actuator has its own stiffness, and the force that it can develop is not constant over its range of movement – as the actuator extends, some force is required to cause its own deformation, and the actuation force available progressively decreases. It is, therefore, important to consider the stiffness of the actuator, which is given by the ratio of the blocking force over the free displacement.

The free displacement or free stroke is the deflection of the actuator in the absence of any opposing force and relates to the maximum material strain, typically 0.1–0.15% for PZT. A lower electric field strength will give a smaller free displacement. The force-generating capability of a piezoelectric actuator is normally quoted as the blocking force. This is the maximum force exerted by an actuator when placed in a perfectly unyielding clamp, which implies that there is *no* displacement of the actuator under these conditions.

Note 4.1

Free displacement is the deflection of the actuator with zero force. The blocking force is the force exerted by the actuator with zero displacement.

Piezoelectric Materials: Properties, Applications and Devices, First Edition.
Christopher R. Bowen and Andrew R. Plummer.

When the maximum free stroke of the actuator is known (with no applied force) and the maximum blocking force is known (with no displacement), the characteristic of the material under combined force generation and deflection is shown using a force-versus-displacement diagram. If x_{free} is the free displacement, F is the actuator force, and x is the actual actuator displacement under combined electrical and mechanical loading, the relationship between them is given by:

$$F = k_{act}\left(x_{free} - x\right) \tag{4.1}$$

where k_{act} is the stiffness of the actuator. Note that x_{free} depends on the electric field strength, i.e. the applied voltage. From Eq. (4.1),

i) when $F = 0$, the displacement x is x_{free}.
ii) when $x = 0$, the force F is the blocking force, i.e.

$$F_{block} = k_{act} x_{free} \tag{4.2}$$

When the piezoelectric actuator is acting against a spring load of stiffness k_{spr}, the opposing force is given by:

$$F = k_{spr} x \tag{4.3}$$

Substituting into Eq. (4.1):

$$k_{spr} x = k_{act}\left(x_{free} - x\right)$$

so

$$x = x_{free}\left(\frac{k_{act}}{k_{act} + k_{spr}}\right) \text{ for a spring load} \tag{4.4}$$

Thus, the reduction in the displacement of the piezoelectric actuator (relative to the free stroke) is a function of the actuator stiffness (k_{act}) and the stiffness of the structure it is acting against (k_{spr}). For example, if $k_{act} = k_{spr}$, exactly one-half of the piezoelectric actuator's free displacement can be generated. Similarly, in this case, the actuator force will also be half the blocking force. Figure 4.1 shows the force–displacement characteristic.

Note 4.2

A piezoelectric actuator will never generate both its blocking force and free displacement simultaneously.

The actuator voltage will be limited by the maximum safe electric field, approximately 2 kV/mm for common varieties of PZT. At lower voltages, both the free displacement and blocking force are reduced; see Figure 4.2. The resulting displacement and force are given by the intersection of the load line with the actuator line for whatever voltage is being used.

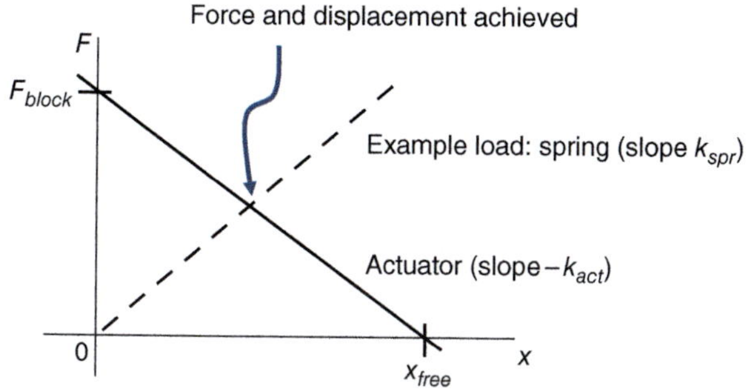

Figure 4.1 Actuator force–displacement characteristic.

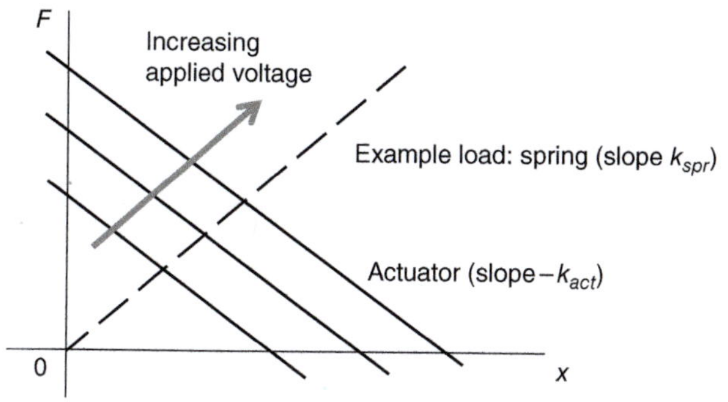

Figure 4.2 Force–displacement characteristic with changing voltage.

4.3 Single-Layer Actuator

A simple actuator is shown in Figure 4.3. A metal electrode is bonded on each end of a block of piezoceramic. Applying a positive voltage V elongates the actuator. Conversely, if V is negative, the actuator shrinks. An external force F will also lead to a deformation due to

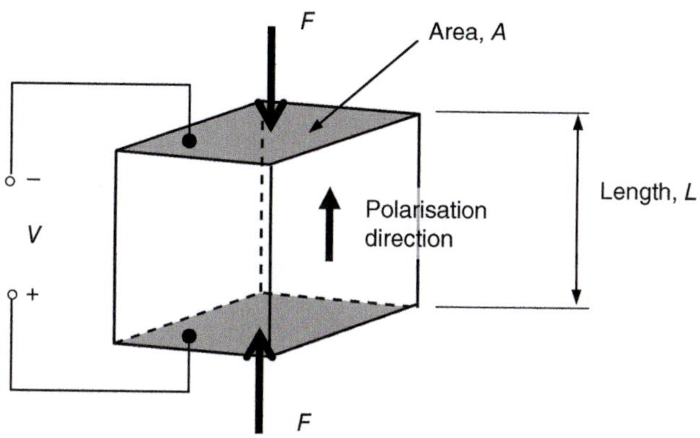

Figure 4.3 Single-layer actuator.

the conventional elastic properties of the material. As shown in Chapter 3, this relationship between extension, force and voltage for an actuator can be derived from the dimension-independent relationship between strain, stress and electric field:

$$S = s\sigma + d\,E_f \tag{4.5}$$

where S is strain, s is the material compliance (measured while short-circuited, i.e. $E_f = 0$), σ is tensile stress, and E_f is the electric field strength. These four quantities are given by:

$$S = \frac{x}{L} \tag{4.6}$$

$$s = \frac{1}{E} \tag{4.7}$$

$$\sigma = -\frac{F}{A} \tag{4.8}$$

$$E_f = \frac{V}{L} \tag{4.9}$$

where x is the extension of the actuator, E is the Young's Modulus of the material, and A and L are defined in Figure 4.3. In Eq. (4.5), d is the piezoelectric charge coefficient or strain coefficient for the particular piezoceramic employed, giving the strain per electric field strength.

Note 4.3

For a single-layer actuator working in the polarisation direction (as shown in Figure 4.3), the material strain coefficient d directly gives the free extension per voltage (m/V).

For a particular actuator of length L and cross-sectional area A, Eq. (4.5) can be expressed as:

$$\frac{x}{L} = -\frac{1}{E}\frac{F}{A} + d\frac{V}{L}$$

$$x = -\frac{L}{EA}F + dV \tag{4.10}$$

where x is the extension of the actuator (i.e. the increase in L). Remember that E is Young's modulus, not the electric field strength (E_f). Comparing to Eq. (4.1), it can be seen that the free displacement, i.e. the displacement with zero force, is given by $x_{free} = dV$, and the blocking force, i.e. the force giving zero displacement, is:

$$F_{block} = \frac{EAd}{L}V \tag{4.11}$$

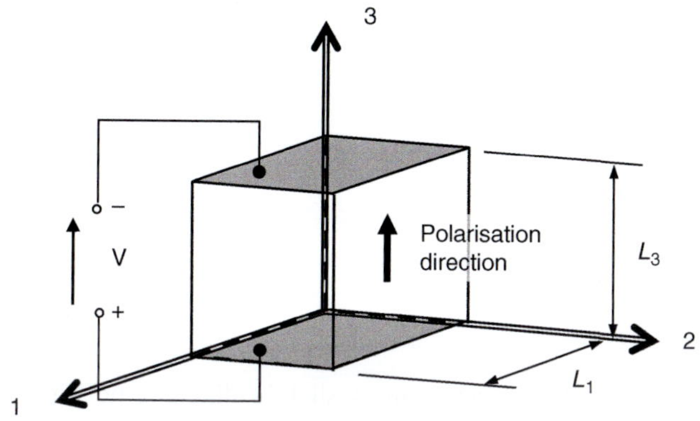

Figure 4.4 Single-layer actuator: axis conventions.

Plotting Eq. (4.10) would give the actuator characteristic shown in Figure 4.2. Note that $\frac{EA}{L}$ is the stiffness of the actuator, k_{act}.

Equation (4.10) calculates the change in length in the direction of polarisation. As length L increases, the area A reduces, so in fact, the other dimensions of the block (width and depth) reduce. In the general case, the block should be considered as a three-dimensional entity, and strain can occur in any or all of those dimensions. The three Cartesian co-ordinates are labelled 1, 2 and 3, as shown in Figure 4.4. If we simply apply an electric field in the polarisation direction, i.e. the 3 direction, Eq. (4.5) can be generalised as:

$$\begin{bmatrix} S_1 \\ S_2 \\ S_3 \end{bmatrix} = \begin{bmatrix} s_{11} & s_{12} & s_{13} \\ s_{12} & s_{11} & s_{13} \\ s_{13} & s_{13} & s_{33} \end{bmatrix} \begin{bmatrix} \sigma_1 \\ \sigma_2 \\ \sigma_3 \end{bmatrix} + \begin{bmatrix} d_{31} \\ d_{31} \\ d_{33} \end{bmatrix} E_{f3} \tag{4.12}$$

Note that this is a portion of Eq. (3.11). In addition, some of the compliance and strain coefficients, s_{ij} and d_{ij}, are repeated as a consequence of symmetries found in common piezoceramic materials.

For an axial actuator, such as a stack, the stress is applied parallel to the field (axis 3), and only the axis 3 strain is important. Alternatively, an actuator may make use of transverse strain, so a stress is experienced perpendicular to the field (e.g. in axis 1), and only the axis 1 strain is important. Applying these conditions reduces Eq. (4.12) to the following scalar equations for the two alternative scenarios, respectively:

Axial:

$$S_3 = s_{33}\sigma_3 + d_{33}E_{f3} \tag{4.13}$$

Transverse:

$$S_1 = s_{11}\sigma_1 + d_{31}E_{f3} \tag{4.14}$$

In terms of extension, force and voltage, Eqs. (4.13) and (4.14) become:

Axial:

$$x_3 = -\frac{L_3}{E_{33}L_1L_2}F_3 + d_{33}V \tag{4.15}$$

Transverse:

$$x_1 = -\frac{L_1}{E_{11}L_2L_3}F_1 + \frac{d_{31}L_1}{L_3}V \tag{4.16}$$

Equation (4.15) is the same as Eq. (4.10), just using a slightly different notation. Equation (4.16) is somewhat different, as the force and extension are perpendicular to the voltage direction. Typical values for Young's modulus (also known as the elastic constant and the reciprocal of compliance) for PZT are 50–70 GPa in the 3 direction (E_{33}) and 60–80 GPa in the 1 direction (E_{11}), measured with the actuator short-circuited. Typical values for PZT strain coefficients are between 300 and 600×10^{-12} m/V for d_{33} and -100 to -300×10^{-12} m/V for d_{31}.

Note 4.4

Actuators can be intended to actuate either parallel or perpendicular to the polarisation direction. The actuator displacement is dependent on the force in the actuation direction and the voltage difference applied in the polarisation direction.

Example: Single-Layer PZT Axial Actuator

An axial actuator is an actuator used to produce a force along the polarisation axis, i.e. Eq. (4.15) applies. A PZT plate has the following specifications:

Dimensions:	$L_1 = 10$ mm, $L_2 = 10$ mm, $L_3 = 0.5$ mm
PZT strain constants:	$d_{31} = -200 \times 10^{-12}$ m/V, $d_{33} = 600 \times 10^{-12}$ m/V
PZT Young's modulus:	$E_{33} = 70 \times 10^9$ N/m^2

With an applied voltage $V = 200$ V, calculate the free displacement and the blocking force, and sketch the graph of force versus displacement.

The term $\frac{L_3}{E_{33}L_1L_2} = \frac{0.0005}{70 \times 10^9 \times 0.01 \times 0.01} = 7.14 \times 10^{-11}$ m/N, which is the compliance of the actuator $1/k_{act}$; therefore, Eq. (4.15) becomes:

$$x_3 = -7.14 \times 10^{-11}F_3 + 6 \times 10^{-10}V$$

Or with $V = 200$ V: $\quad x_3 = -7.14 \times 10^{-11}F_3 + 1.2 \times 10^{-7}$

Therefore, the free displacement ($F_3 = 0$) is 0.12 μm, and the blocking force ($x_3 = 0$) is 1680 N.

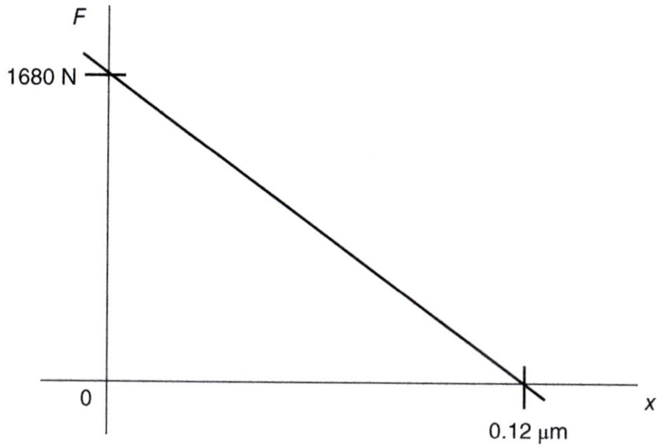

Figure 4.5 Example force–displacement characteristic for a single-layer actuator at 200 V.

Figure 4.5 shows the force–displacement characteristic at this applied voltage. Note that this is unlikely to be the limit of performance, as the field strength is only 0.4 kV/mm, whereas the limit for PZT is typically around 2 kV/mm. Thus, we might expect to be able to use voltages up to about 1000 V in this example.

4.4 Stack Actuators

The maximum strain of a piezoceramic is typically in the region of 0.1–0.15% along axis 3, achieved with a field strength of around 2 kV/mm in the polarisation direction for PZT. This gives rise to an expansion in axis 3. A reverse electric field at about 30% of this strength is possible before depolarisation occurs, giving a maximum contraction of around 0.04%.

The maximum displacement (stroke) of an axial actuator, utilising axis 3 strain, is proportional to its height (L_3). A convenient way to increase the height is to stack a number of thin piezoceramic discs or plates forming a column, interleaving them with electrodes, as shown in Figure 4.6. As the electrode spacing can remain small, the voltage does not have to be higher to achieve the required electric field strength.

The force–displacement–voltage relationship of Eq. (4.10) still applies, except that $d_{33}V$ gives the free displacement of a single layer, which must be scaled by the number of layers, n:

$$x = -\frac{L}{E_{33}A}F + nd_{33}V \tag{4.17}$$

Adding more layers of the same thickness increases both n and L; therefore, although the free displacement increases, the blocking force does not. This can be seen by rearranging Eq. (4.17):

$$F_{block} = \frac{E_{33}An}{L}d_{33}V \tag{4.18}$$

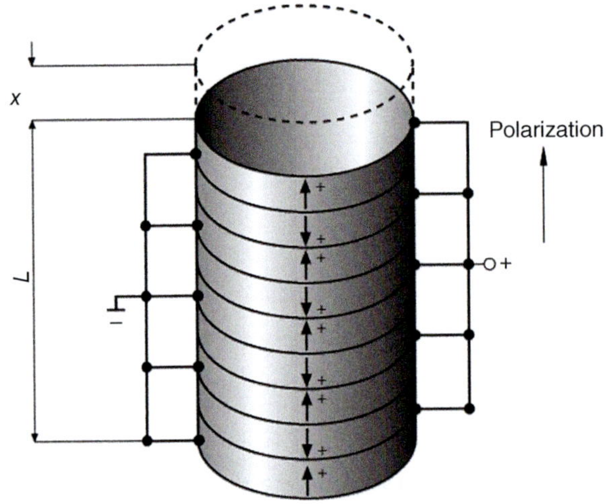

Figure 4.6 A stack actuator. *Source:* Courtesy of Physik Instrumente [PI] GmbH & Co. KG.

Or

$$F_{block} = \frac{E_{33}A}{h} d_{33}V \tag{4.19}$$

where h is the thickness of one layer. Note that the electrode thickness can usually be considered negligible compared to the thickness of the piezoceramic. To increase the blocking force, the cross-sectional area A of the actuator would need to be increased (hence increasing the actuator stiffness).

4.4.1 Actuator Preloading

The free displacement of a stack is not significantly influenced by a constant load (such as a preload or a weight). A static load merely provides an initial offset of the zero-point position dependent on the load level and actuator stiffness.

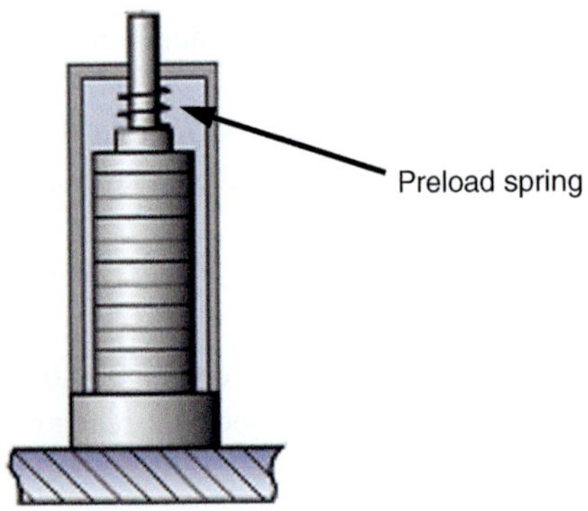

Figure 4.7 Preloaded stack actuator. *Source:* Courtesy of Physik Instrumente [PI] GmbH & Co. KG.

Piezoelectric ceramics should not be loaded in tension, and to avoid static tensile loads, stack actuators are often placed in a state of compression; see Figure 4.7. A preload of up to 50% of the blocking force is generally suitable. While the free displacement of the actuator is reduced when acting against a load of finite stiffness (whereby the restraining force increases with deflection), the free displacement is not significantly influenced by a constant load, although an excessive force could lead to depolarisation. Thus, the preload spring shown in Figure 4.7 must be an order of magnitude less stiff than the actuator itself.

4.4.2 Piezoelectric Stack Actuator Selection Example

Actuator performance (displacement and force generation) for actuators of different blocking forces and free stroke can be compared by plotting force–displacement diagrams. This will be undertaken based on some examples of commercially available piezoelectric stack actuators. The specific example requirement is to provide as large a displacement as possible while working against a spring of stiffness 100 N/μm. In this example, only extension is considered.

Note that the price of a stack depends on the manufacturing process. Monolithic multilayer actuators, which tend to have a lower cross-sectional area, can be manufactured in the same way as ceramic capacitors and are of lower cost than assembled stack actuators. Larger stack actuators consist of individually assembled thin piezoceramic elements and are labour-intensive for manufacturing.

Table 4.1 shows examples of large stack actuators from Noliac and Physik Instrumente. The greater the length of the actuator, the more the free stroke. The blocking forces shown in Table 4.1 are a function of free stroke and actuator stiffness ($F_{block} = k_{act}x_{free}$).

The SCMA-R5 and P-025.200 stacks have large free strokes of 282 μm and 300 μm, respectively, due to their length (200 mm and 240 mm); this is in general agreement with the maximum strain being of the order of 0.1–0.15%. The blocking force and stiffness are smaller for the SCMA-R5 stack than for the P-025.200 stack. The actuator lengths (L) and material elastic modulus (E) are similar; therefore, the differences in blocking force and stiffness are primarily due to the smaller diameter and cross-sectional area of the SCMA-R5 ring stack, which leads to lower stiffness.

Figure 4.8 shows a force-versus-displacement plot for the SCMA-R5 ring stack, which exhibits a free displacement of 282 μm and a blocking force of 8 kN at the maximum operating voltage of 200 V. The solid line is the actuator characteristic, and the axis intercepts are the extremes of performance (free displacement and blocking force). To consider the actuator performance when acting against a spring of stiffness $k_{spr} = 100$ N/μm, the spring load line (dashed) is also shown; the gradient of the line is the spring stiffness. The intercept of the two lines is the maximum displacement and force that could be achieved (indicated with arrows). Due to the low stiffness of this actuator ($k_{act} = 28$ N/μm) compared to the 100-N/μm spring, it can be seen that the achievable displacement of the actuator is much less than the free stroke of 282 μm. The possible expansive displacement is only 62 μm, and the force generated is 6.2 kN. The use of this actuator highlights the low displacement when $k_{spr} >> k_{act}$, as would be expected from Eq. (4.4).

Figure 4.9 shows a force-versus-displacement plot for the P-025.200 actuator with a free displacement of 300 μm and a higher blocking force of 16 kN at the maximum operating voltage of 1000 V. The larger diameter and corresponding higher stiffness of 58 N/μm (compared to the SCMA-R5 stack) lead to a smaller drop in possible displacement when acting against a spring of stiffness 100 N/μm, but k_{act} is still lower than k_{spr}. In this case, this possible displacement is 105 μm with a force generation of 10.5 kN.

Since the actuator must deliver a large displacement, the actuator length must be relatively large. However, the stiffness decreases as the stack gets longer, which is why $k_{act} \ll k_{spr}$ for the two actuators considered in Figures 4.8 and 4.9. One way to increase the stiffness of a large stroke, long-length actuator would be to increase the cross-sectional

Table 4.1 Example large piezoelectric stack actuators from Noliac (first row) and PI (rows 2 and 3).

Actuator	Free stroke (μm)	Diameter (mm)	Length, L (mm)	Blocking force (kN)	Stiffness (N/μm)	Free strain (free stroke/ length) (%)	Capacitance (μF)	Operating voltage (V)	Maximum Temperature (°C)	Current (A) at 20 Hz[a]	Estimated mass (kg)[b]	ΔT (°C)[c]	Resonant Frequency (kHz)
SCMA-R5 ring stack	282	20 OD 12 ID	200	8.0	28	0.14	77.6	0–200	150	1.0 A at 200 V	0.17	~4	>6
P-025.200 cylindrical stack	300	25	244	16	54	0.12	8.5	0–1000	85	0.52 A at 1000 V	0.9	~7.5	5
P-056.90P cylindrical stack	180	56	169	70	390	0.11	21	0–1000	150	1.3 A at 1000 V	3.12	~17	7

[a] the maximum current is found from frequency, capacitance and peak-to-peak voltage using πfCV; see Eq. (3.31).
[b] mass given by the product of volume and density; assuming the density of PZT is 7500 kg m^{-3} and ignoring electrode material.
[c] Equation (3.28) used to provide an estimate of temperature increase based on $f = 20$ Hz, $u = 2$ kJ/m^3 [1] and $k(T)$ of 30 $Wm^{-2}K^{-1}$ (which is a suitable value for a freely suspended stack).

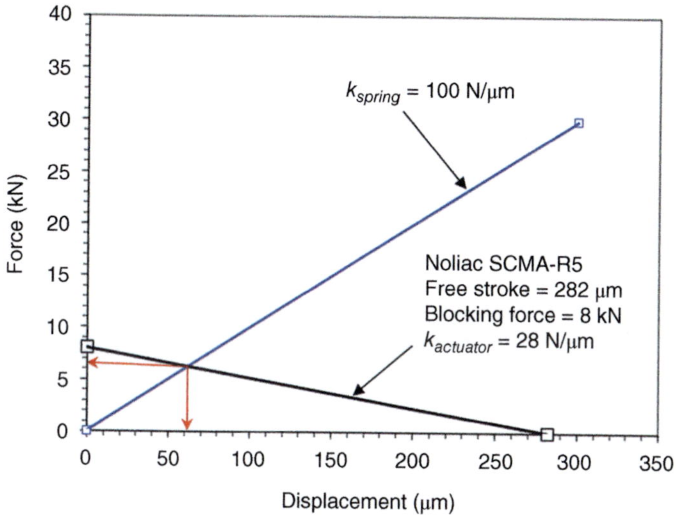

Figure 4.8 Force versus displacement for SCMA-R5 stack.

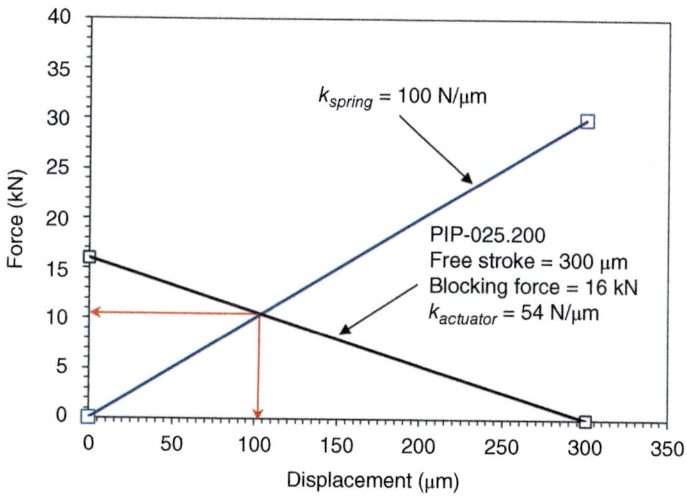

Figure 4.9 Force generation versus displacement for P-025.200 stack.

area of the actuator (noting that Young's modulus does not change significantly between different PZT types). Stack P-056.90P has a large diameter, 56 mm, but it is shorter and provides only a 180-μm free stroke. The large area and reduced length provide a high stiffness (390 N/μm) and high blocking force (70 kN). Figure 4.10 shows that although the free displacement is small compared to the previous actuators, the achievable displacement acting against the spring is in excess of 140 μm with a force of 14 kN, and so is the best of the three actuators.

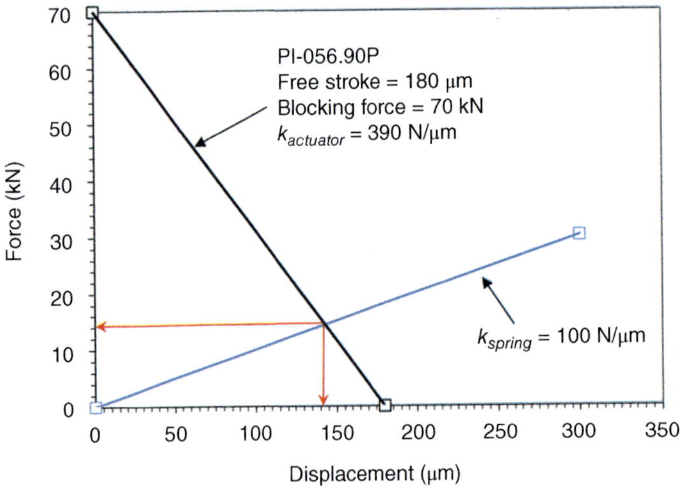

Figure 4.10 Force generation versus displacement for P-056.90 stack.

4.4.3 Optimum Stack Dimensions

More generally, it can be shown that the best stack dimensions when acting against a spring are those which make the actuator stiffness the same as the spring stiffness. This is shown below.

The actuator force–displacement characteristic, Eq. (4.17), can be rewritten for the maximum voltage condition as:

$$x = -\frac{L}{E_{33}A}F + S_{free}L \tag{4.20}$$

where S_{free} is the free strain in the axial direction at maximum voltage (typically between 0.1% and 0.15%). This can also be expressed in terms of the actuator stiffness $k_{act} = \frac{E_{33}A}{L}$ and the volume of the piezoceramic $V_p = AL$:

$$x = -\frac{1}{k_{act}}F + S_{free}\sqrt{\frac{E_{33}V_p}{k_{act}}} \tag{4.21}$$

When acting against a spring,

$$F = k_{spr}x$$

so:

$$x = S_{free}\sqrt{E_{33}V_p}\frac{\sqrt{k_{act}}}{k_{act} + k_{spr}} \tag{4.22}$$

To find the maximum displacement which can be achieved for a fixed stack volume, Eq. (4.22) can be used to find the derivative of x with respect to k_{act}, and this derivative can be set to zero:

$$\frac{dx}{dk_{act}} = S_{free}\sqrt{E_{33}V_p}\left(\frac{1}{2\sqrt{k_{act}}\left(k_{act} + k_{spr}\right)} - \frac{\sqrt{k_{act}}}{\left(k_{act} + k_{spr}\right)^2}\right) = 0 \tag{4.23}$$

This has the solution $k_{act} = k_{spr}$, and thus, from Eq. (4.22), the maximum displacement for that volume of the piezoceramic is:

$$x = \frac{S_{free}}{2}\sqrt{\frac{E_{33}V_p}{k_{act}}} = \frac{S_{free}L}{2} = \frac{x_{free}}{2} \tag{4.24}$$

Note 4.5

When acting against a spring, the best stack in terms of the most efficient use of piezoceramic material has a stiffness which is the same as the spring stiffness.

4.4.4 Piezoelectric Actuator Stack Sizing Guidelines

Often, an amplification mechanism is required to gear the output of a piezoceramic stack, typically to increase its range of movement. Thus, as the gearing ratio can be changed, the stack displacement and force ranges are not directly dictated by the application requirement, but the amount of actuator work output is (i.e. the product of force and displacement).

As shown in Figure 4.11, the force–displacement plot can be redrawn as a stress–strain plot. The slope is related to Young's modulus, which is 50–70 GN/m^2 for PZT, but the stiffness is usually lower in a multilayer stack (around 30 GN/m^2). As shown, a typical operating envelope is up to half the blocking force and half the free displacement; the stack can generate any combination of forces and displacements in that range.

The work density of an actuator is the maximum work output per unit volume:

$$W_\rho = \frac{\frac{F_{block}}{2}\cdot\frac{x_{free}}{2}}{AL} = \frac{\sigma_{block}S_{free}}{4} \approx 10\ \text{mJ/cm}^3 \tag{4.25}$$

Thus, knowing the work output, the actuator volume can be estimated. If the gear ratio is different so that the required force and displacement range map onto a different rectangle in Figure 4.11, then the actuator volume required is larger. For example, if only a quarter of the free displacement is used, then three-quarters of the blocking force is available, and the work density calculation of Eq. (4.25) gives a lower value, 7.5 mJ/cm^3.

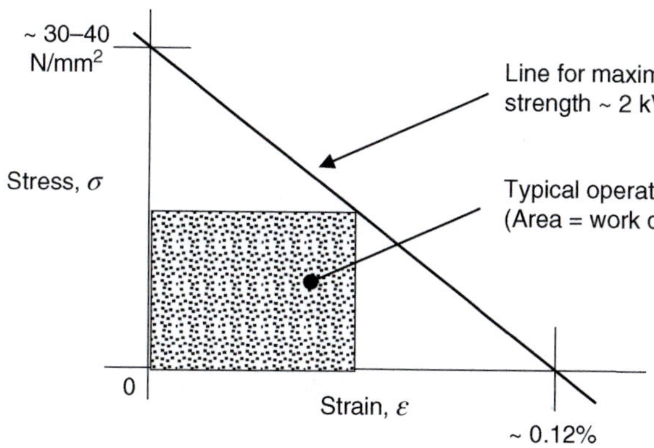

Figure 4.11 Generalised piezoelectric stack stress–strain generation.

Note 4.6

Stack sizing (or gearing) to maximise work output will use about half the blocking force and half the free displacement, giving ~10 mJ/cm^3.

Note 4.7

Bending actuators (Section 4.5) have a lower work density, ~2 mJ/cm^3.

The electrical input required can also be estimated. The maximum electric field is similar for many stacks, but the maximum actuator voltage can be reduced by making the layers thinner (but increasing their number). This increases the capacitance of the actuator considerably (a square law), and thus, the current increases in dynamic operation. However, the product of current and voltage remains constant.

Applying a voltage V to a stack will increase its stored charge Q; the stored energy is given by:

$$E_{elec} = \frac{VQ}{2} = \frac{CV^2}{2} \tag{4.26}$$

The actuator capacitance C is related to the number of layers n (as there are effectively a number of capacitors in parallel), electrode area, electrode separation (L/n) and the permittivity of the ceramic:

$$C = n\frac{\varepsilon_{33}^{\sigma} A}{L/n} = \frac{\varepsilon_{33}^{\sigma} A n^2}{L} \tag{4.27}$$

$\varepsilon_{33}^{\sigma}$ is the permittivity for zero stress; thus, we are assuming a zero-load extension. Substituting for C in Eq. (4.26) and noting that V is the product of the electric field strength E_f and the layer thickness (L/n), the maximum electrical energy per unit volume is:

$$\frac{E_{elec}}{AL} = \frac{\varepsilon_{33}^{\sigma} {E_f}^2}{2} \approx 36\,\mathrm{mJ/cm^3}$$

given that the maximum $E_f \approx 2\,\mathrm{kV/mm}$ and $\varepsilon_{33}^{\sigma} \approx 2000\varepsilon_o = 1.8 \times 10^{-8}$ F/m. As described in Section 3.2.3 in relation to coupling coefficients, around 50% of this energy could be converted into mechanical work, but given that the typical working range (Figure 4.11) will not capture the full area under the force–displacement curve, ~10 mJ/cm^3 is a likely practical maximum mechanical work output as previously established.

As the electrical energy per unit volume is a constant, the product of charge and voltage is a constant (Eq. (4.23)). This demonstrates that reducing the voltage required by having many thinner layers will increase the current requirement.

Note 4.8

The electrical energy required to cycle a stack is ~36 mJ/cm^3, compared to a maximum of ~10 mJ/cm^3 converted to mechanical energy, and ~ 2 mJ/cm^3 lost in hysteresis (Section 3.13).

4.5 Rectangular Bending Actuators (Bimorphs)

4.5.1 Bimorph Characteristics

Traditionally, a rectangular bimorph consists of two piezoceramic strips bonded together, electrically connected so that one extends while the other contracts, thus bending the actuator (Figure 4.12). This has the disadvantage that the direction of the electric field applied to one strip is always opposite to the poling direction, for example, the top strip in the figure, when bending downwards, as shown. A three-wire multilayer bending actuator overcomes this issue by biasing the field such that at its mid-position, layers in each half of the bender experience half the maximum field strength. Commercially available actuators of this type often have many thin layers to allow a relatively stiff, high-force characteristic whilst only requiring a modest voltage (e.g. 100 V peak). A cross-section of such an actuator is shown in Figure 4.13.

Referring to Figure 4.13, one set of electrodes is maintained at a negative voltage ($-V_s$), one set is maintained at a positive voltage ($+V_s$), and the control voltage V is connected to the intervening electrodes and varies between $-V_s$ and $+V_s$. This electrode configuration allows deflection of the actuator in both directions since half of the piezoceramic layers expand in-plane, while the other half contracts as the control voltage varies from zero to $-V_s$ or from zero to $+V_s$. When the control voltage is zero, the upper and lower half layers experience the same electric field, resulting in no bending displacement.

The multilayer bimorph can be modelled in the same way as two piezoceramic strips, each of thickness $h/2$, bonded together, as shown in Figure 4.12 [2]. In reality, each of

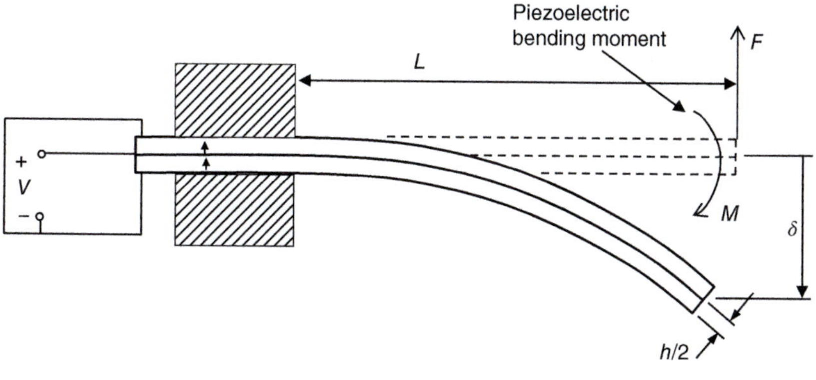

Figure 4.12 Bimorph bending.

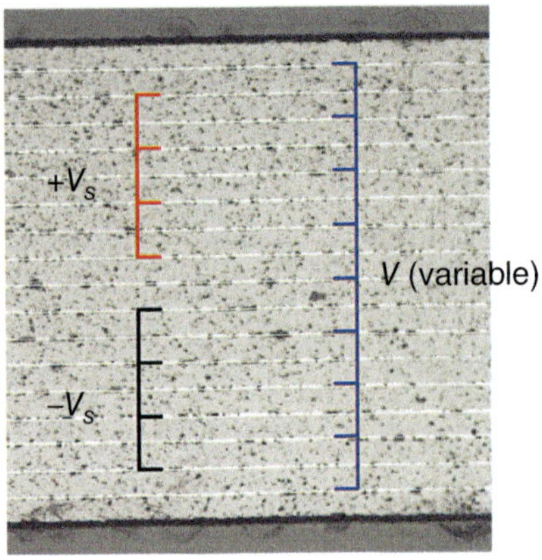

Figure 4.13 Cross-section through multilayer bimorph (18 piezoceramic layers separated by 17 electrodes, giving 16 active layers). Courtesy of Johan Persson.

the strips is composed of multiple layers. The thickness of the internal electrodes is assumed to be negligible in the following analysis. The bimorph generates a piezoelectric moment, M, when subjected to a control voltage, V. This moment can be related to the blocking force, F_b, of each strip and hence the blocking stress σ_{1b} of the piezoceramic, thus:

$$M = F_b h/2$$

or

$$M = \sigma_{1b} b h^2/4 \tag{4.28}$$

where b is the bimorph width. The blocking stress can be expressed in terms of electric field strength, E_{f3}, Young's modulus, E_{11}, and strain constant, d_{31} by

$$\sigma_{1b} = -d_{31} E_{f3} E_{11} \tag{4.29}$$

This is derived from Eq. (4.14). Substituting into Eq. (4.28) gives

$$M = -d_{31} E_{f3} E_{11} b h^2/4 \tag{4.30}$$

For a three-wire multilayer bimorph, E_{f3} is actually the deviation in field strength away from that present in the undeformed position obtained when the control voltage V is zero. The electric field strength depends on the thickness of each layer.

$$E_{f3} = \frac{V}{\frac{h}{n_l}} = \frac{n_l V}{h} \tag{4.31}$$

where n_l is the number of piezoceramic layers, and V is the applied voltage. This gives

$$M = -d_{31} n_l E_{11} b h V/4 \tag{4.32}$$

The bending moment is related to the curvature of a beam ($1/R$, where R is the radius shown in Figure 4.14):

$$\frac{1}{R} = \frac{M}{E_{11} I} \tag{4.33}$$

The second moment of area, I_b, for a bimorph of thickness h and width b is given by:

$$I = \frac{bh^3}{12} \tag{4.34}$$

Substituting M and I into Eq. (4.33) using Eqs. (4.32) and (4.34), respectively:

$$\frac{1}{R} = -3 \frac{d_{31} n_l}{h^2} V \tag{4.35}$$

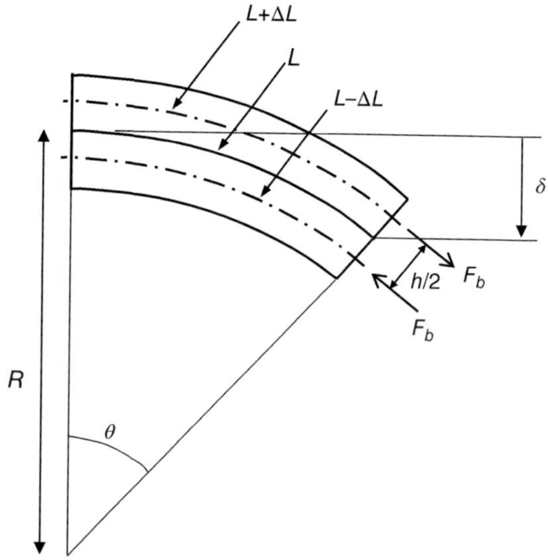

Figure 4.14 Relationship between curvature and displacement for a small bimorph beam element.

Remember that d_{31} is negative, so for the way we have set up the analysis, a positive control voltage gives a positive moment and downward deflection in Figure 4.12.

It remains to be determined how the deflection and curvature are related. Referring to Figure 4.14:

$$\delta = R(1 - \cos\theta) \tag{4.36}$$

The angle θ will be small, so the approximation $\cos\theta \approx 1 - \frac{\theta^2}{2}$ can be used:

$$\delta = \frac{R\theta^2}{2} \tag{4.37}$$

and $\theta = L/R$, so:

$$\delta = \frac{L^2}{2R} \tag{4.38}$$

Substituting Eq. (4.35) into Eq. (4.38):

$$\delta = -\frac{3L^2 d_{31} n_l}{2h^2} V \tag{4.39}$$

Hence, this is the deflection due to the converse piezoelectric effect. Any external forces will also contribute to the deflection, and these contributions can be determined from standard formulae for the deflection of a cantilever beam. If there is an external force F acting on the bimorph in addition to the piezoelectric moment, as shown in Figure 4.12, the deflection is given by:

$$\delta = -\frac{3L^2 d_{31} n_l}{2h^2} V - \frac{L^3}{3E_{11} I} F \tag{4.40}$$

or

$$\delta = -\frac{3L^2 d_{31} n_l}{2h^2} V - \frac{4L^3}{E_{11} b h^3} F \tag{4.41}$$

The stroke of a bimorph actuator is considerably more than a single piezoceramic element or a typical stack actuator. However, inevitably, the force output is proportionately less.

Example: PZT Rectangular Bending Actuator (Bimorph)
A multilayer PZT bimorph has 10 active layers (n_l) and the following parameters:

Dimensions:	$L = 20$ mm, $b = 5$ mm, thickness $h = 0.5$ mm
PZT strain constants:	$d_{31} = -200 \times 10^{-12}$ m/V, $d_{33} = 600 \times 10^{-12}$ m/V
PZT Young's modulus:	$E_{11} = 70 \times 10^9$ N/m^2

With a voltage $V = 10$ V, calculate:

a) The free extension (i.e. when the force F is zero)
b) The blocked force (i.e. when the deflection is zero)

Solution.
Displacement and blocking force are governed by Eq. (4.41):

$$\delta = -\frac{3L^2 d_{31} n_l}{2h^2} V - \frac{4L^3}{E_{11} b h^3} F = c_1 V - c_2 F$$

With the length/height ratio $\frac{L}{h} = \frac{20}{0.5} = 40$, we get:

$$c_1 = -\frac{3}{2} d_{31} n_l \left(\frac{L}{h}\right)^2 = 9.6 \times 10^{-7} \text{m/V}$$

$$c_2 = \frac{4}{E_{11} b} \left(\frac{L}{h}\right)^3 = 7.31 \times 10^4 \text{m/N}$$

So the free extension is:

$$F = 0 \Rightarrow \delta = 9.6 \times 10^{-5} \text{m} = 96\ \mu\text{m}$$

The blocking force is:

$$\delta = 0 \Rightarrow F = \frac{9.6 \times 10^{-5}}{7.31 \times 10^{-4}} \text{N} = 0.131\ \text{N}$$

Thus, the force–displacement characteristic is shown in Figure 4.15.

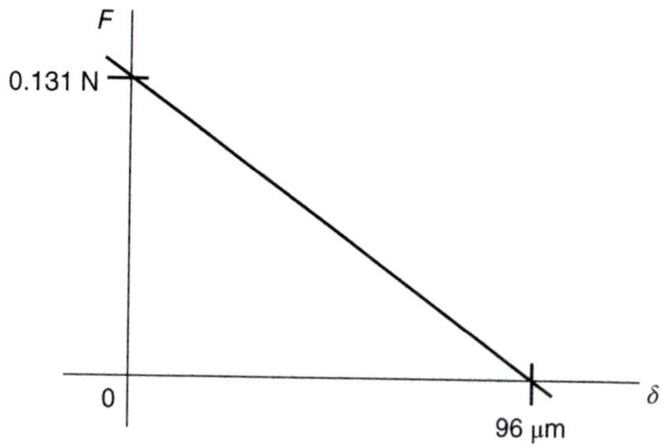

Figure 4.15 Example force–displacement characteristic for rectangular bimorph.

In summary, we can calculate the force–displacement characteristics of rectangular bimorphs. The actuator configuration provides a relatively straightforward way to increase the displacement in comparison to a stack actuator, although inevitably with a reduction in force capability. It should be noted that by utilising the d_{31} strain coefficient, which is typically only about a third of the value of d_{33}, the work output per volume of the piezoceramic is reduced. One other point about the calculation method: it assumes that all the piezoceramic is active. The actuator cross-section in Figure 4.13 shows that the top and bottom surfaces are inactive ceramic, i.e. the layers are not between electrodes. Although the effect is likely to be quite small, this fact could be taken into account in the equation derivations.

Unlike stack actuators, some piezoceramic layers will inevitably be in tension. Typically, tensile strength is significantly lower than compressive strength, and failure by cracking is a danger [3]. For example, a PZT may have a compressive strength of 250 MPa (although depoling is a danger at around 30% of this value) but a tensile strength of about 40 MPa [4]. Failure modes and in situ health monitoring of bending actuators are discussed further in [5].

Note 4.9

For a similar actuator length, bending actuators have much higher strokes than stack actuators, but at the expense of much lower force output and reduced work density.

4.5.2 Other Rectangular Benders

A number of variants of the rectangular bender concept exist, and a notable example is the THUNDER actuator: THin Layer UNimorph ferroelectric DrivER and Sensor, originally patented by NASA [6]. It is composed of a thin wafer of PZT bonded to an electrically conductive substrate and a superstrate with high-performance epoxy (Figure 4.16). With just one PZT layer, it can only bend in one direction (hence, unimorph), rather than two directions like a bimorph.

(a)

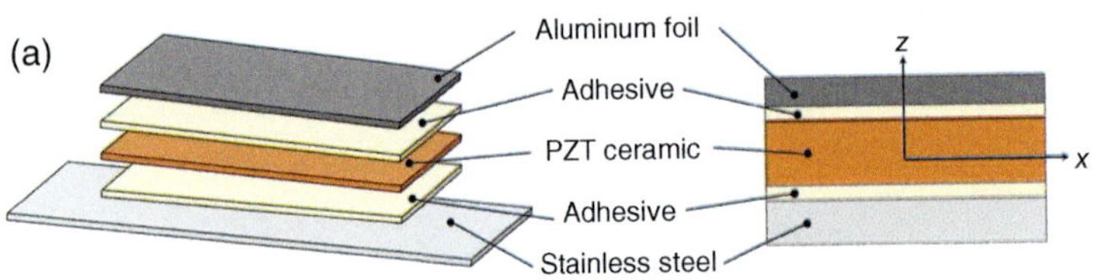

(b)

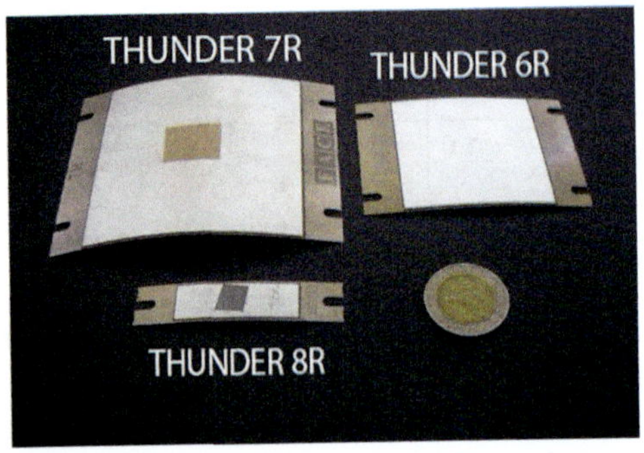

Figure 4.16 THUNDER unimorph actuator; (a) layout of the laminated actuator, (b) examples of THUNDER devices (6R, 7R and 8R). *Source:* [6] / MDPI / CC BY 4.0.

4.6 Ring Benders

Ring benders only differ in shape from rectangular benders; in other respects, their construction is the same. Thus, they are formed from a number of thin piezoceramic layers separated by electrodes, which are normally grouped in three sets, as shown in Figure 4.13. The flat annular disc deforms with a non-zero variable voltage, forming either a convex or a concave shape depending on the voltage polarity (Figure 4.17). Unlike a rectangular bender, a ring bender is able to harness the piezoelectric strain in both in-plane directions, i.e. due to both d_{31} and d_{32}.

4.6.1 Ring Bender Deformation Analysis

For pure compound bending of a thin piezoelectric bimorph, the bending moment at any point is related to curvature (both in and out of plane) and the piezoelectric bending moment M_e:

$$M_1 = D\left(\frac{1}{\rho_1} + \mu\frac{1}{\rho_2}\right) - M_e \tag{4.42}$$

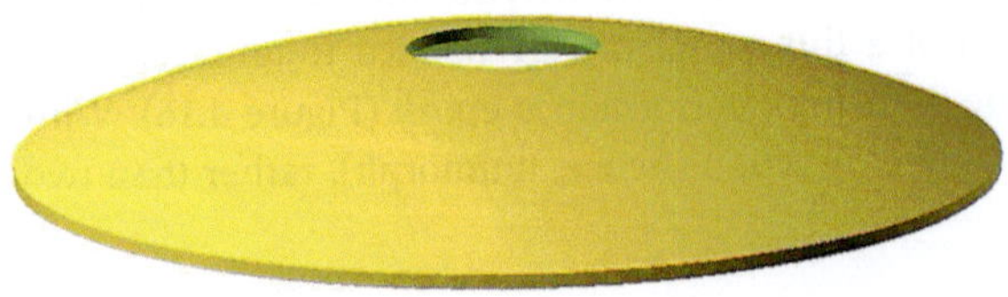

Figure 4.17 Ring bender showing a deformed shape with non-zero driving voltage.

where M_1 is the bending moment per unit length, $1/\rho_1$ is the curvature in the same plane, $1/\rho_2$ is the out-of-plane curvature, μ is Poisson's ratio, and D is the flexural rigidity per unit length:

$$D = \frac{E_{11}h^3}{12(1-\mu^2)}$$

where E_{11} is the transverse Young's modulus, and h is the bimorph thickness [7]. This assumes pure bending, i.e. the lateral sides of any small element cut from the thin plate remain plane and rotate about an axis within the middle plane (neutral surface) of the plate.

The in-plane blocking force per unit length of material for the top half of the piezoelectric ring is given by:

$$F_e = d_{31}E_{f3}E_{11}\frac{h_a}{2}$$

where d_{31} $(= d_{32})$ is the piezoelectric strain coefficient orthogonal to the poling direction, h_a is the active thickness, and E_{f3} is the field strength. The two different thicknesses, h_a and h, account for the fact that not all the piezoceramic may be subject to an electric field, i.e. there may be outer inactive layers (see Figure 4.13). An equal and opposite blocking force will be generated on the bottom half of the disc. The points of action of the two forces will be separated by distance $h_a/2$ and so equated to a moment $F_e h_a/2$.

A small element in the ring is shown in Figure 4.18, together with the blocking force in the top layer (half). Equal and opposite forces exist in the bottom layer. Consider the equivalent blocking moments in this two-dimensional structure. By taking moments about the tangential direction, the additional moment at the outer radius $(r + \Delta r)$ compared to the inner radius is due to the piezoelectric force difference in the radial direction, as well as a component from the circumferential direction:

$$M_e \Delta r \Delta\theta = (F_e(r + \Delta r)\Delta\theta - F_e r \Delta\theta + F_e \Delta r \Delta\theta)\frac{h_a}{2}$$

$$M_e = F_e h_a$$

$$M_e = d_{31}E_{f3}E_{11}\frac{{h_a}^2}{4}$$

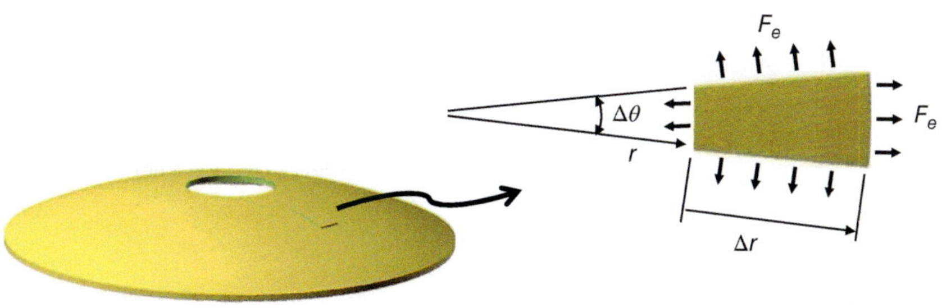

Figure 4.18 Effective piezoelectric forces (top layer).

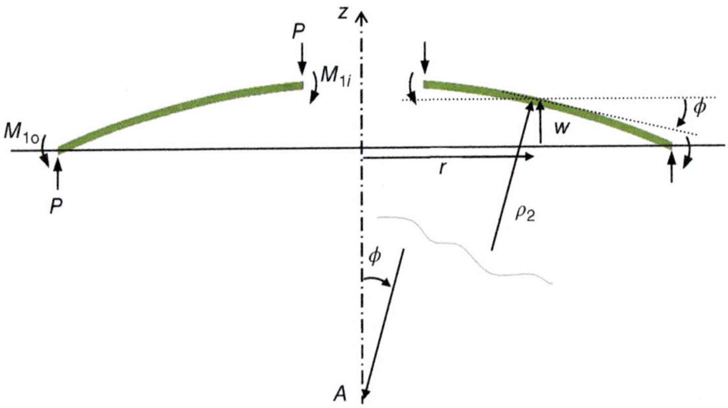

Figure 4.19 Cross-section through ring bender.

As shown in Figure 4.19, for a symmetrically loaded disc or ring, if $1/\rho_1$ is the curvature at some radius r along a diametral section, then:

$$\frac{1}{\rho_1} = \frac{d\phi}{dr}$$

where ϕ is the slope (or angle) at that point on the section, recognising that a circumferential cross-section at radius r forms a conical surface with apex A, the orthogonal curvature is:

$$\frac{1}{\rho_2} = \frac{\phi}{r}$$

Thus, from (4.42):

$$M_1 = D\left(\frac{d\phi}{dr} + \mu\frac{\phi}{r}\right) - M_e \tag{4.43}$$

$$M_2 = D\left(\frac{\phi}{r} + \mu\frac{d\phi}{dr}\right) - M_e \tag{4.44}$$

Consider the small element shown in Figure 4.18. Figure 4.20 is a side view showing the moment and shear force per unit length acting on the element. Moment M_2 acts on the sides of length Δr and has a component in the plane shown of $M_2\Delta r\Delta\theta$. Hence, summing moments gives:

$$(M_1 + \Delta M_1)(r + \Delta r)\Delta\theta = M_1 r\Delta\theta + V_f r\Delta\theta\Delta r + M_2\Delta r\Delta\theta$$

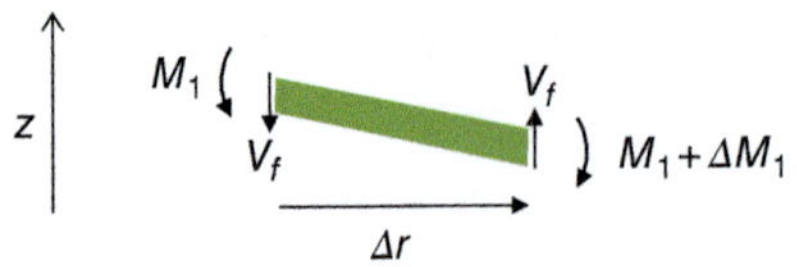

Figure 4.20 Cross-section of small element.

Neglecting small higher-order terms, in the limit, this becomes:

$$M_1 + \frac{dM_1}{dr} r = V_f r + M_2$$

Substituting using Eqs. (4.43) and (4.44):

$$\frac{d^2\phi}{dr^2} + \frac{1}{r}\frac{d\phi}{dr} - \frac{\phi}{r^2} = \frac{V_f}{D}$$

or

$$\frac{d}{dr}\left[\frac{1}{r}\frac{d(x\phi)}{dr}\right] = \frac{V_f}{D}$$

The ring bender exerts a force P, as shown in Figure 4.19, so the shear force per unit length is:

$$V_f = \frac{P}{2\pi r}$$

giving

$$\frac{d}{dr}\left[\frac{1}{r}\frac{d(r\phi)}{dr}\right] = \frac{P}{2\pi D r}$$

This can be integrated twice to give:

$$\phi = \frac{Pr}{8\pi D}(2\ \ln r - 1) + \frac{C_1 r}{2} + \frac{C_2}{r}$$

Redefining constants of integration C_1 and C_2 to simplify the subsequent derivation:

$$\phi = -\frac{1}{8\pi D}\left[-Pr\left(1 + 2\ln\frac{r}{r_o}\right) + 2C_3 r + \frac{C_4 {r_o}^2}{r}\right] \tag{4.45}$$

The deflection w of the ring bender can be found by integration from:

$$\phi = -\frac{dw}{dr}$$

giving

$$w = \frac{{r_o}^2}{8\pi D}\left[-P\frac{r^2}{{r_o}^2}\ln\frac{r}{r_o} + C_3\frac{r^2}{{r_o}^2} + C_4 \ln r\right] + C_5$$

The inner edge displacement (at radius r_i) relative to the outer edge is given by:

$$w_i = \frac{{r_o}^2}{8\pi D}\left[-P\frac{{r_i}^2}{{r_o}^2}\ln\frac{r_i}{r_o} + C_3\frac{{r_i}^2}{{r_o}^2} + C_4 \ln r_i + P\ln 1 - C_3 - C_4 \ln r_o\right]$$

or

$$w_i = \frac{r_o^{\,2}}{8\pi D}\left[-PR^2 \ln R - C_3\left(1 - R^2\right) + C_4 \ln R\right] \tag{4.46}$$

where the ratio of inner to outer radii is $R = r_i/r_o$.

The constants of integration can be found from the boundary conditions. At the outer edge, the moment is given by the product of the angle and torsional stiffness of the mounting:

$$M_{1o} = -k_o\phi_o$$

Substituting into Eq. (4.43) and rearranging:

$$8\pi M_e = 8\pi D\left.\frac{d\phi}{dr}\right|_o + 8\pi D\mu_o\frac{\phi_o}{r_o} \tag{4.47}$$

where

$$\mu_o = \mu + \frac{r_o k_o}{D}$$

Differentiating Eq. (4.45) gives:

$$\frac{d\phi}{dr} = -\frac{1}{8\pi D}\left[-P\left(3 + 2\ln\frac{r}{r_o}\right) + 2C_3 - \frac{C_4 r_o^2}{r^2}\right] \tag{4.48}$$

Substituting Eqs. (4.45) and (4.48) into Eq. (4.47):

$$-8\pi M_e = (-3P + 2C_3 - C_4) + \mu_o(-P + 2C_3 + C_4) \tag{4.49}$$

Similarly, at the inner edge, the moment is given by the product of the angle and torsional stiffness of the mounting:

$$M_{1i} = k_i\phi_i$$

Substituting into Eq. (4.43) and rearranging:

$$8\pi M_e = 8\pi D\left.\frac{d\phi}{dr}\right|_i + 8\pi D\mu_i\frac{\phi_i}{r_i} \tag{4.50}$$

where

$$\mu_i = \mu - \frac{r_i k_i}{D}$$

Substituting Eqs. (4.45) and (4.48) into Eq. (4.50):

$$-8\pi M_e = \left[-P(3 + 2\ln R) + 2C_3 - \frac{C_4}{R^2}\right] + \mu_i\left[-P(1 + 2\ln R) + 2C_3 + \frac{C_4}{R^2}\right] \tag{4.51}$$

The constants can be calculated from Eqs. (4.49) and (4.51), which can be rearranged into the following matrix equation:

$$\begin{bmatrix} -8\pi M_e + P(\mu_o + 3) \\ -8\pi M_e + P(\mu_i + 3 + 2\ln R(\mu_i + 1)) \end{bmatrix} = \begin{bmatrix} 2(\mu_o + 1) & (\mu_o - 1) \\ 2(\mu_i + 1) & \frac{1}{R^2}(\mu_i - 1) \end{bmatrix} \begin{bmatrix} C_3 \\ C_4 \end{bmatrix} \tag{4.52}$$

Note that this derivation assumes the neutral surface is strain free. If built in at both edges, this will give an error, but this will be small due to the small deflections expected for ring benders.

Example: PZT Ring Bender

A multilayer PZT ring bender is analysed. The device is a Noliac CMBR07 bender, with properties given below.

Dimensions	Outer radius, $r_o = 20\,\text{mm}$
	Inner radius, $r_i = 4\,\text{mm}$
	Thickness, $h = 0.67\,\text{mm}$
	Active thickness, $h_a = 0.57\,\text{mm}$
	Number of layers, $n = 8$
Material properties	Young's modulus, $E_{11} = 59\,\text{GN/m}^2$
	Poisson's ratio, $\mu = 0.3$
	Strain constant, $d_{31} = -170 \times 10^{-12}\,\text{m/V}$
Loading/driving condition	Nominal blocking force, $P_b = 13\,\text{N}$
	Maximum drive voltage, $V = 200\,\text{V}$
	Gives field strength, $E_f = \dfrac{V}{h_a/n} = 2.8\,\text{kV/mm}$

The displacement w can be determined as a function of radius, which is plotted in Figures 4.21–4.24 for the applied load and voltage given in the table. The radius is normalised so that the outer edge is 1.0, and thus, in this case, the inner edge is 0.2. Four cases are considered for different combinations of inner and outer edge torsional mounting stiffnesses (k_i and k_o, respectively):

1) $k_i = 0$ and $k_o = 0$
2) $k_i \approx \infty$ and $k_o = 0$
3) $k_i = 0$ and $k_o \approx \infty$
4) $k_i \approx \infty$ and $k_o \approx \infty$

These results illustrate that the edge-mounting torsional stiffness affects displacement, and in particular, it is important to have compliance at the outer edge to allow the actuator to move appropriately. Nevertheless, a practical actuator will require the ring bender to have some form of inner and outer edge fixation to be able to exert an axial load force, and so careful design is required to prevent excessive movement restriction.

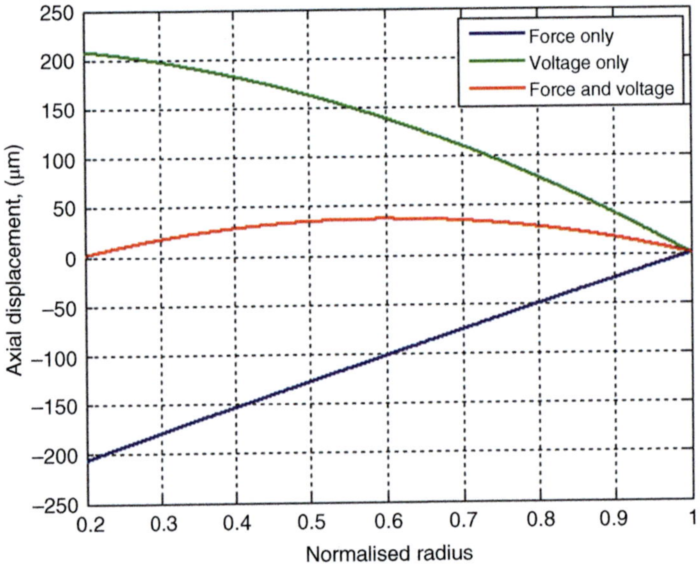

Figure 4.21 Case 1 – zero mounting stiffness.

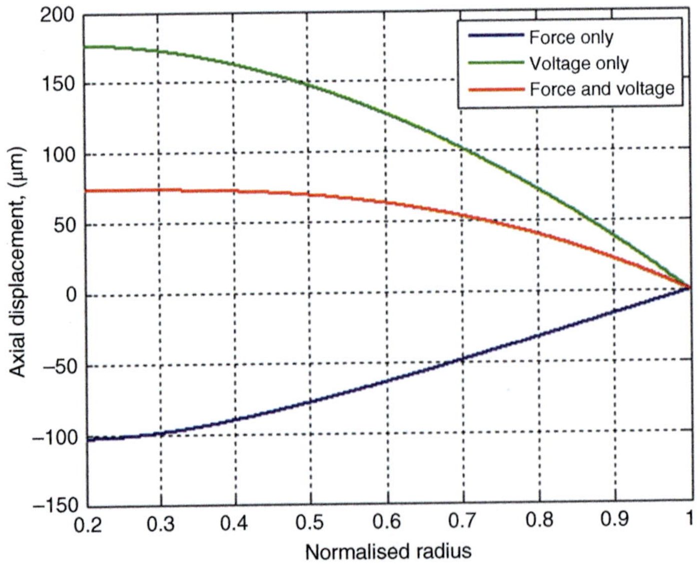

Figure 4.22 Case 2 – very high inner edge stiffness (≈ rigid).

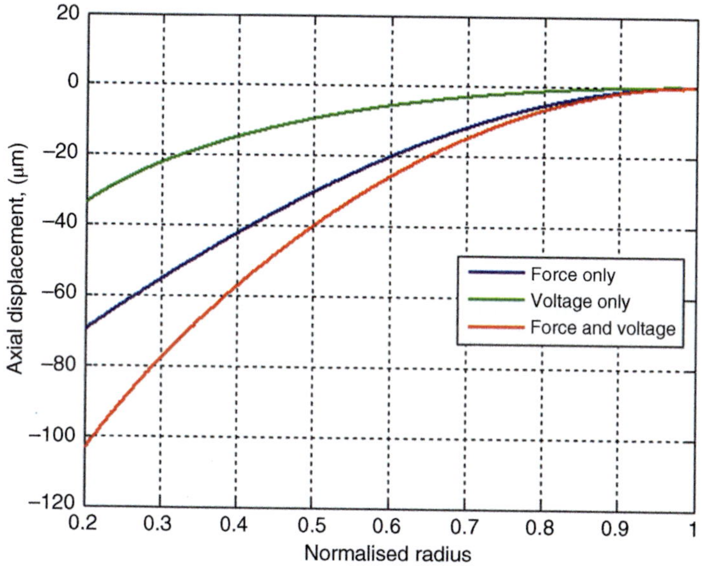

Figure 4.23 Case 3 – very high outer edge stiffness (≈ rigid).

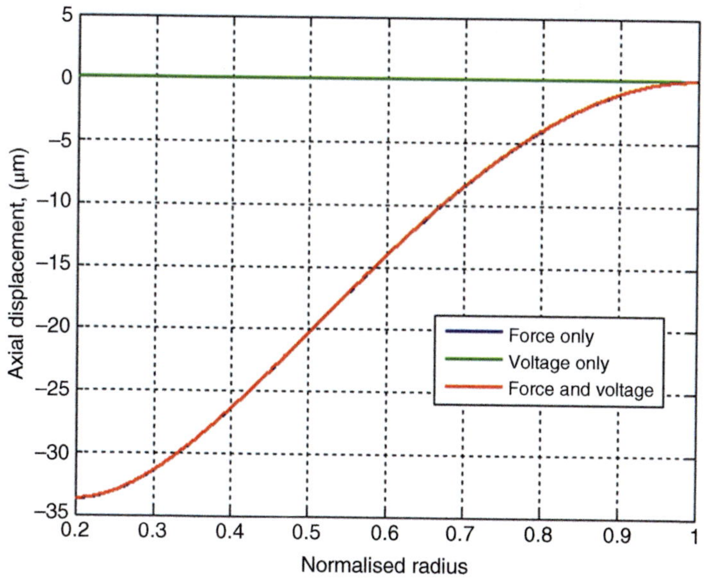

Figure 4.24 Case 4 – very high stiffness at both edges.

Note 4.10

Outer edge mounting compliance is a key parameter. A very stiff outer edge mount will actually cause the ring bender to deflect in the opposite direction compared to a freely moving actuator.

4.6.2 Ring Bender Free Displacement and Blocking Force

Here, we restrict ourselves to just considering the axial displacement of the inner edge relative to the outer edge (i.e. free displacement) and the axial force generated if the inner and outer edges are constrained to remain in the same plane (i.e. the blocking force). The torsional stiffnesses of inner and outer edge mountings are still included.

The free displacement is achieved when the load, P, is zero. From Eq. (4.46):

$$w_{if} = \frac{{r_o}^2}{8\pi D}\left[-C_3\left(1-R^2\right)+C_4\ln R\right]$$

and from Eq. (4.52)

$$-8\pi M_e\begin{bmatrix}1\\1\end{bmatrix} = \begin{bmatrix}2(\mu_o+1) & (\mu_o-1)\\ 2(\mu_i+1) & \frac{1}{R^2}(\mu_i-1)\end{bmatrix}\begin{bmatrix}C_3\\C_4\end{bmatrix}$$

giving free displacement:

$$w_{if} = \frac{M_e {r_o}^2}{D}\left[(1-R^2)-\ln R\right]\begin{bmatrix}2(\mu_o+1) & (\mu_o-1)\\ 2(\mu_i+1) & \frac{1}{R^2}(\mu_i-1)\end{bmatrix}^{-1}\begin{bmatrix}1\\1\end{bmatrix}$$

The force which prevents movement, i.e. $w_i = 0$, is the blocking force. From Eq. (4.46):

$$P_b = -C_3\frac{(1-R^2)}{R^2\ln R} + C_4\frac{1}{R^2}$$

Substituting for P in Eq. (4.52):

$$-8\pi M_e\begin{bmatrix}1\\1\end{bmatrix} + \frac{1}{R^2}\begin{bmatrix}-\frac{(1-R^2)}{\ln R}(\mu_o+3) & (\mu_o+3)\\ -\frac{(1-R^2)}{\ln R}(\mu_i+3+2(\mu_i+1)\ln R) & (\mu_i+3+2(\mu_i+1)\ln R)\end{bmatrix}\begin{bmatrix}C_3\\C_4\end{bmatrix}$$
$$= \begin{bmatrix}2(\mu_o+1) & (\mu_o-1)\\ 2(\mu_i+1) & \frac{1}{R^2}(\mu_i-1)\end{bmatrix}\begin{bmatrix}C_3\\C_4\end{bmatrix}$$

or

$$-8\pi R^2 M_e \begin{bmatrix}1\\1\end{bmatrix} = \begin{bmatrix} 2R^2(\mu_o+1)+\dfrac{(1-R^2)}{\ln R}(\mu_o+3) & R^2(\mu_o-1)-(\mu_o+3) \\ 2R^2(\mu_i+1)+\dfrac{(1-R^2)}{\ln R}(\mu_i+3+2(\mu_i+1)\ln R) & -2(2+(\mu_i+1)\ln R) \end{bmatrix} \begin{bmatrix}C_3\\C_4\end{bmatrix}$$

from which C_3 and C_4 and hence P_b can be found.

Contour plots for varying mounting stiffnesses are shown in Figures 4.25 and 4.26 for the previous PZT ring bender example. The normalised inner and outer edge stiffnesses are $\frac{r_i k_i}{D}$ and $\frac{r_o k_o}{D}$, respectively, which are also the same as $\mu - \mu_i$ and $\mu_o - \mu$, respectively. Note that the shape of the curves is only influenced by the value of R, with a change in μ just causing a shift along both axes, and other parameters just affecting the magnitude of the free displacement or blocking force.

The force–displacement characteristic of the actuator is a linear relationship joining the free displacement and blocking force conditions, in the same way as for other piezoelectric actuators (e.g. Figure 4.15). The area under the graph is the maximum work output of the actuator:

$$\frac{w_{if} P_b}{2}$$

Thus, from the data in Figures 4.25 and 4.26, the maximum work output can also be plotted, as shown in Figure 4.27. This shows that the torsional outer edge stiffness has a big impact on work output. The work output is less sensitive to the inner edge stiffness, but actually, stiffening the inner edge mount does tend to increase the work output to some degree.

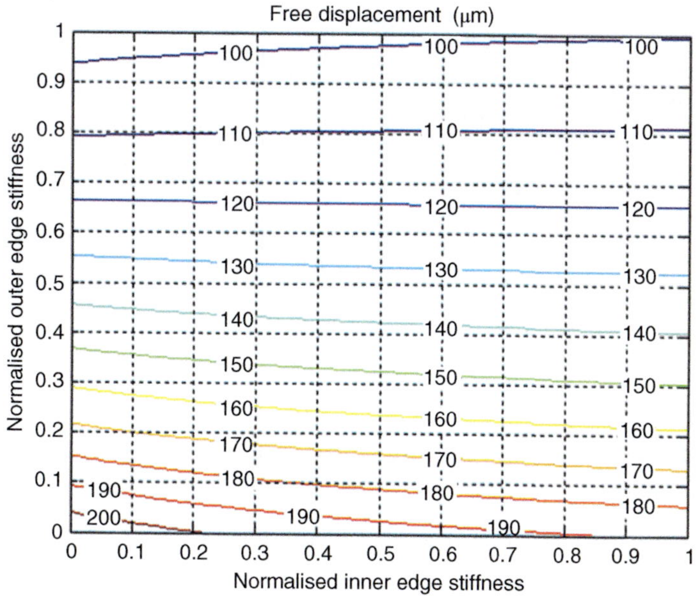

Figure 4.25 Free displacement for a range of mounting torsional stiffnesses.

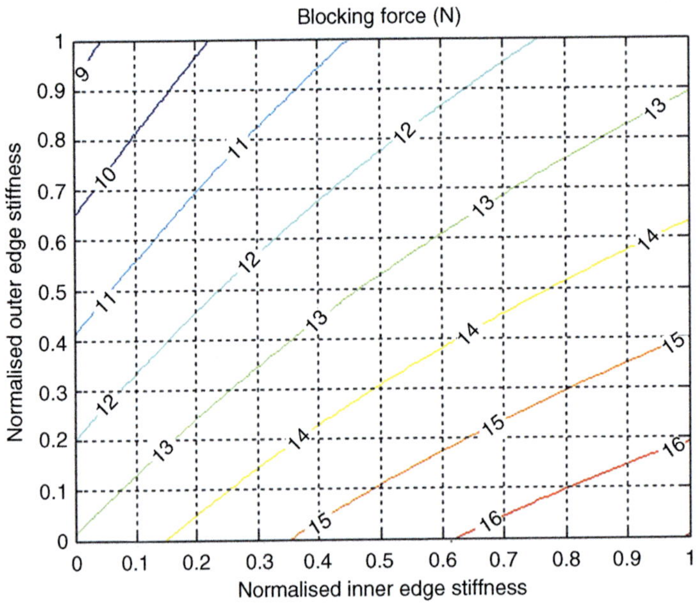

Figure 4.26 Blocking force for a range of mounting torsional stiffnesses.

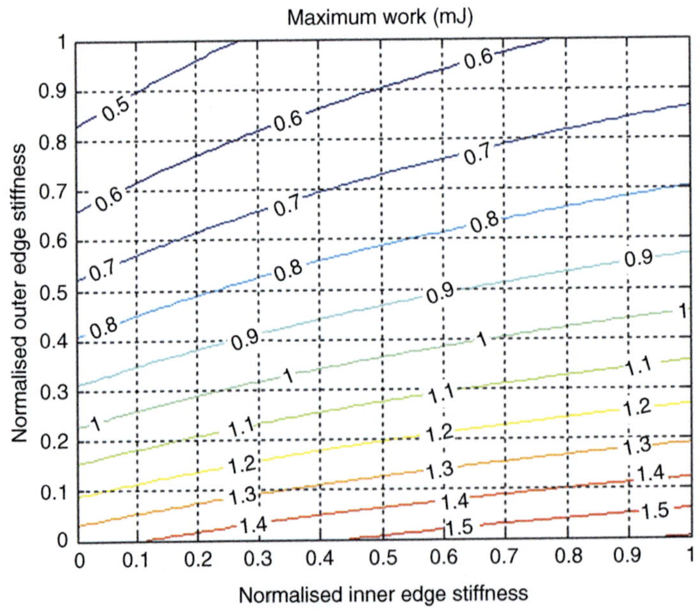

Figure 4.27 Maximum work output for a range of mounting torsional stiffnesses.

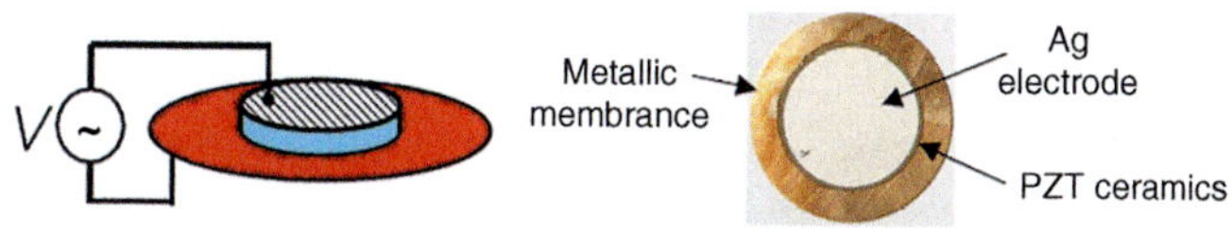

Figure 4.28 Circular unimorph. Courtesy of APC International Ltd.

Note 4.11
Ring benders usually have a higher force output than rectangular benders of similar area and thickness.

4.6.3 Other Circular Benders

Small, low-cost unimorph circular benders are also available, typically with a single PZT disc bonded to a metal diaphragm (Figure 4.28); they are a circular equivalent to the THUNDER actuator. They are used as simple sound generators, often in alarms or buzzers.

4.7 Mechanical Amplification

Various mechanical arrangements are possible to amplify the displacement range of a stack. This, by necessity, will reduce the force output. A key challenge is to remove any backlash in the mechanism, as this is otherwise likely to dominate compared to the movement range of the stack. Thus, flexures are preferred rather than traditional rotating joints. Another challenge is thermal sensitivity. To ensure that thermal expansion or contraction of the stack is not picked up and amplified, the mechanism structure that bridges the stack must be matched to the stack, i.e. have the same thermal expansion coefficient as far as possible. A third challenge is that for effective dynamic operation, the mass of the mechanism must be small and of high stiffness to ensure that resonant frequencies remain above the operational frequency range.

The flextensional actuator has proved to be an effective amplification concept. The stack is embedded in a diamond-shaped metal spring, which provides a compressive stack preload; see Figure 4.29. The dimensions of the diamond, specifically the length ratio of the major axis (in line with the stack) to the minor axis, give the amplification ratio.

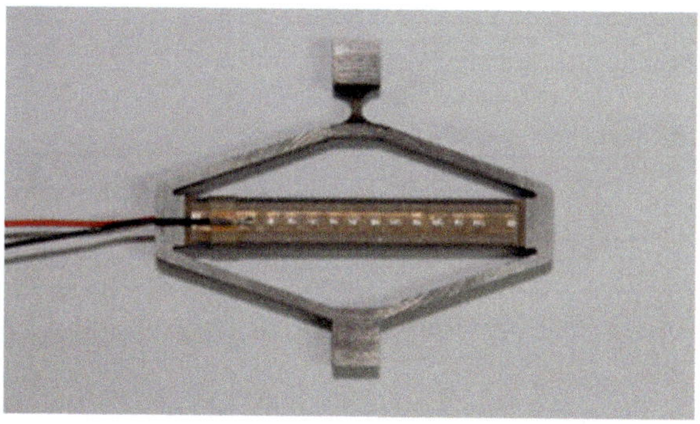

Figure 4.29 Flextensional amplified stack actuator. Horizontal expansion of the stack creates a larger vertical contraction, which is used as the actuator output. *Source:* [8] / with permission of Elsevier.

Flextensional actuators fit between bimorphs and unamplified stacks in terms of displacement and force capabilities. Note that the output motion is perpendicular to the stack, across the two apexes on the minor axis, so typically one apex is fixed, and the whole stack will move sideways with half the displacement of the actuator. This characteristic and the resonant modes of the device more generally limit the dynamic response.

Table 4.2 provides the details of some example commercial flextensional actuators and compares them with an electromagnetic voice coil actuator of similar force output. Note

Table 4.2 Example commercial flextensional stack actuators and comparison with an electromagnetic voice coil actuator. Courtesy of Cedrat Technologies, DSM and Sensata.

	Cedrat piezo 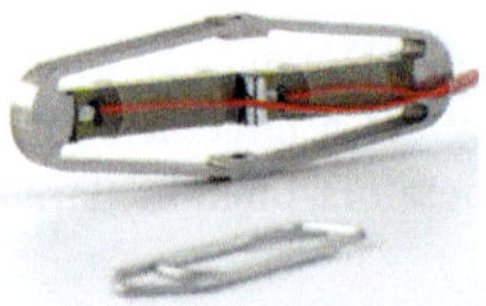	DSM piezo 	Sensata voice coil 
Basic specifications	**APA 200 M v 3.3** Stroke: 230 μm Blocked force: 73 N Maximum voltage: 150 V Capacitance: 3.15 μF Stiffness: 0.32 N/μm Effective moving mass: 10 g Natural frequency: 900 Hz Natural frequency with a 75-g load: 309 Hz	**FPA-0200E-S-0518-150-SS** Stroke: 200 μm Blocked force: 100 N Maximum voltage: 150 V Capacitance: 1.6 μF Stiffness: 0.53 N/μm Effective moving mass: 24 g Natural frequency: 750 Hz Natural frequency with a 75-g load: 368 Hz	**LAS16-23-000A** Stroke: 6.36 mm Peak force: 89 N Continuous stall force: 24.5 N Peak current: 7.02 A Continuous current: 1.93 A Peak voltage: 33 V DC resistance: 4.7 Ω Inductance: 1.25 μH Weight of coil: 51.7 g
Dimensions	Width × Height × Depth: 53.2 mm × 17 mm × 9 mm	Width × Height × Depth: 28 mm × 15 mm × 10 mm	Diameter × Height: Ø 40 mm × 58.42 mm
Mounting	Precision mounting with allowance for fine-tuning is necessary.		Mounts through faceplate.
Temperature sensitivity	Strain decreases in the order of 2 μm/°C for environments below 15 °C.		Resistance increases with temperature but otherwise has minimal effects.
Self-heating	Maximum frequency at full amplitude around 200 Hz is recommended.		The maximum frequency is 470 Hz for a 10-s burst or 250 Hz continuous.
Control	Low damping means resonance is easily excited. Harder to control around a full stroke due to lack of force.		Relatively linear. Position control requires velocity feedback to improve damping.
Response time	0.94 ms with a 5-A amplifier.	0.96 ms with a 2-A amplifier.	1.13 ms.

Figure 4.30 Miniature flextensional disc actuators, (a) Moonie structure, (b) Cymbal structure. *Source:* [9] / with permission of Elsevier.

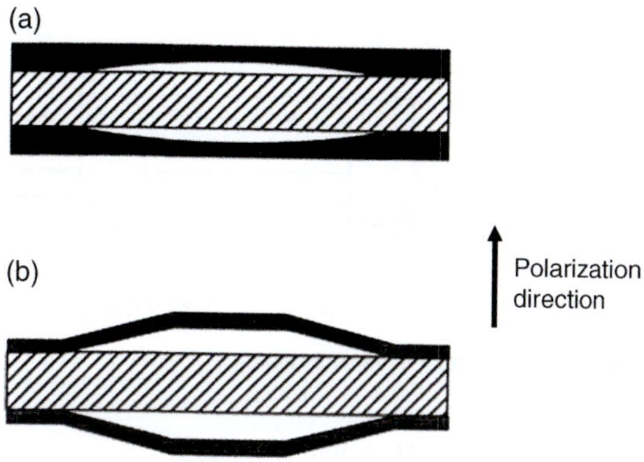

Note 4.12

Mechanical amplification mechanisms need high joint stiffness (no backlash), low mass to give high resonant frequencies and matched thermal expansion.

that the effective moving mass of the piezoelectric actuators is substantially smaller than that of the voice coil actuator. The voice coil actuator has a much larger stroke, but its peak force output is not sustainable due to self-heating, and the continuous force output is much lower than that of the piezoelectric actuators.

The flextensional concept can be applied to single-layer piezoceramic discs also. The 'Moonie' or 'Cymbal' structures shown in Figure 4.30 have metal end caps which bow as a result of contraction in the transverse (1) direction, thus adding to the expansion in the axial (3) direction.

4.8 Complex Actuator Design

The fundamental ability to harness piezoelectric deformation in axial or flexural actuators has been discussed in previous sections. However, a number of more complex and structured actuator designs have proved their viability, often incorporating several piezoelectric elements. Some examples are briefly described here.

4.8.1 Motion Accumulation

The small displacement range of a piezoelectric actuator, particularly a stack, can be overcome by accumulating the motion of an oscillating actuator. In an inchworm motor, this is achieved by mechanical means. Figure 4.31 shows the concept. Three stack actuators are used, one to provide linear displacement and the other two to clamp and release the ends of the actuator so that travel only occurs in the desired direction.

A piezoelectric pump (or just 'piezo pump') is a hydraulic means of motion accumulation. A single-cylinder pump can be created using a piston driven by a stack or a bending

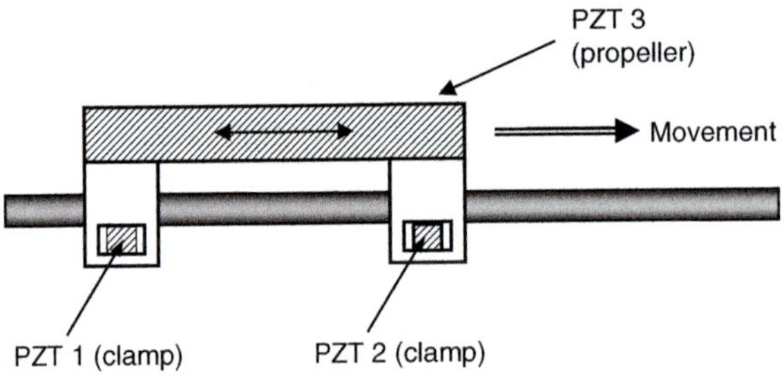

Operating sequence: 1 clamps, 2 releases, 3 extends, 2 clamps, 1 releases, 3 retracts, 1clamps etc.

Figure 4.31 Inchworm motor.

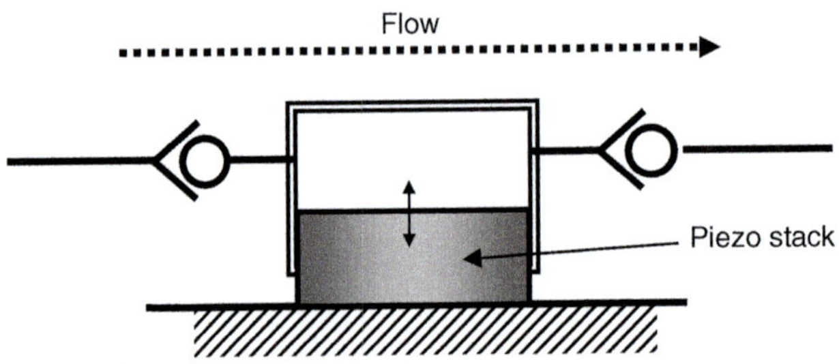

Figure 4.32 Piezoelectric pump.

piezoelectric actuator where low-pressure fluctuations are sufficient. Non-return valves can be used to ensure the flow is unidirectional, as shown in Figure 4.32. The flow can be directed to a hydraulic cylinder if output motion is required. As an alternative to passive non-return valves, actuated valves can be used, driven in synchrony with the piezoelectric actuator. By changing the phasing, the direction of flow can be reversed. Piezo pumps with active valves typically also use piezoelectric valve actuation. One field where piezo pumps have been successfully used, driven by disc actuators, is for controlled drug delivery, requiring very low output power (low flow rate and output pressure).

4.8.2 Ultrasonic Motors

An ultrasonic motor is a rotary piezoelectric actuator, which uses piezoelectric elements driven at a high frequency (e.g. 40 kHz) to resonate a stator ring. This creates a standing wave around the ring. By careful control of the supply voltage, the standing wave can be made to rotate either clockwise or anticlockwise. The stator ring presses against a rotor, and the rotation of the standing wave turns the rotor. This is shown in Figure 4.33. The effect relies on friction between rotor and stator, so the rotor has a hard-wearing, high-friction coating.

Ultrasonic motors have the following desirable characteristics:

i) high torque and low speed
ii) high holding torque when de-energised
iii) fast response
iv) compactness – especially small axial dimension (length)

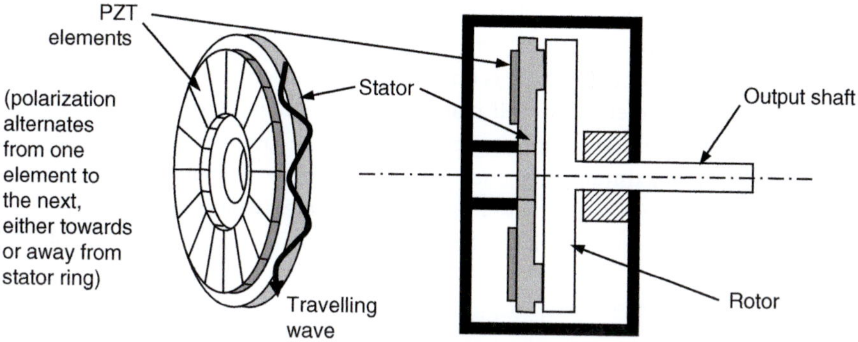

Figure 4.33 Ultrasonic motor.

They are, however, expensive compared to conventional electromagnetic motors. Devices using ultrasonic motors include autofocus cameras, in which they are used for turning the focusing ring.

Another form of ultrasonic motor is shown in Figure 4.34. Here, a tooth is driven to prescribe a cyclic motion at high frequency, typically 100 kHz. The tooth bears against a slider, which consequently moves via the intermittent impetus provided, resulting in a linear motion of up to 0.5 m/s and 15 N force. The drive relies on friction and can resist external loads when unpowered.

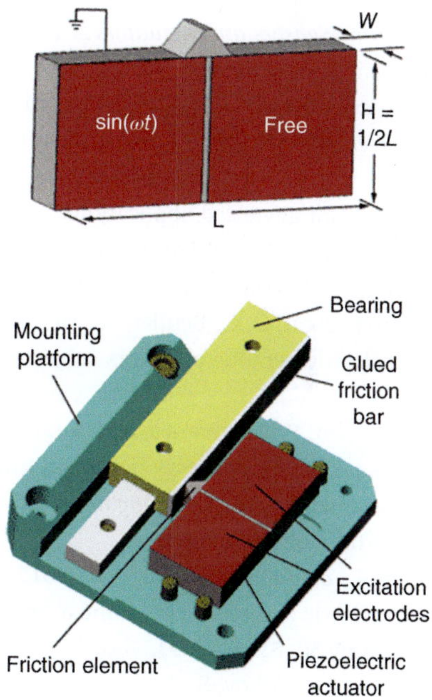

Figure 4.34 PILine® ultrasonic ceramic motor. *Source:* Courtesy of Physik Instrumente [PI] GmbH & Co. KG.

4.9 Concluding Remarks

Piezoelectric ceramics provide a mechanism for converting electrical energy to mechanical work, which is entirely different from conventional actuators using electromagnetic principles. Axial actuators, using d_{33}-mode strain, are straightforward in concept; to achieve useable displacements, they usually use a stack of piezoceramic slices to increase length without the need for higher maximum voltage. Flexural actuators make use of transverse d_{13}-mode strain to bend out of plane, naturally increasing stroke at the expense of force; modelling their force–displacement relationship is more complex than for stack actuators. Providing adequate mechanical mounting and load connections without constraining actuator movement can be a challenge.

For many engineering applications, stacks do not give adequate displacement, and mechanically amplified stacks using various concepts to gear up the displacement are available. Motion accumulation is an alternative approach, making use of the ability of piezoceramics to oscillate at high frequency, e.g. a piezo pump driving a hydraulic cylinder.

Whatever actuator design is employed, it is important for the user to ensure that load displacement and force requirements are well matched to the actuator. At the full actuator displacement (or speed in the case of motion accumulation), the force which can be exerted diminishes to zero, and at maximum force, the displacement is zero. In both cases, the work output of the actuator becomes zero; a better balance is observed where the load displacement and force requirements are well within the free displacement and blocking force of the actuator.

References

1 Pritchard, J., Ramesh, R., and Bowen, C.R. (2004). Time–temperature profiles of multi-layer actuators. *Sensors and Actuators A* 115: 140–145.

2 Sangiah, D.K., Plummer, A.R., Bowen, C.R., and Guerrier, P. (2013). A novel piezohydraulic aerospace servovalve. Part 1: design and modelling. *Proceedings of the Institution of Mechanical Engineers, Part I: Journal of Systems and Control Engineering.* 227 (4): 371–389.

3 Li, T., Chen, Y.H., and Ma, J. (2009). Factors affecting the performance of piezoelectric bending actuators for advanced applications: an overview. *Journal of Materials Science* 44: 5393–5407.

4 Fett, T., Munz, D., and Thun, G. (1999). Tensile and bending strength of piezoelectric ceramics. *Journal of Materials Science Letters* 18: 1899–1902.

5 Aimiyekagbon, O.K., Bender, A., Hemsel, T., and Sextro, W. (2024). Diagnostics of piezoelectric bending actuators subjected to varying operating conditions. *Electronics* 13 (3): 521.

6 Thonapalin, P., Aimmanee, S., Laoratanakul, P., and Das, R. (2021). Thermomechanical effects on electrical energy harvested from laminated piezoelectric devices. *Crystals* 11: 141.

7 Timoshenko, S. (1956). *Strength of Materials, Part II: Advanced Theory and Problems*, 3rde. New York: Van Nostrand.

8 Chen, J., Zhang, C., Xu, M. et al. (2015). Rhombic micro-displacement amplifier for piezoelectric actuator and its linear and hybrid model. *Mechanical Systems and Signal Processing* 50–51: 580–593.

9 Tressler, J.F., Alkoy, S., Dogan, A., and Newnham, R.E. (1999). Functional composites for sensors, actuators and transducers. *Composites Part A: Applied Science and Manufacturing* 30: 477–482.

5

Sensors

5.1 Introduction

A *sensor* is a device that measures a physical quantity by converting it into an equivalent electrical signal. Sensors are often called *transducers* to acknowledge the 'transformation' that they perform. There are a number of possible methods for measurement of a number of important physical quantities, which include acceleration, force, pressure and strain, and there is a wide variety of sensors in industrial use. This chapter provides an overview of the application of piezoelectric materials as sensors for the sensing of force, pressure, acceleration and acoustic sound navigation and ranging (sonar) signals when used in hydrophones. Electrical circuits for amplifying sensor output signals are discussed in Chapter 7, along with filtering circuits for attenuating errors such as high-frequency measurement noise or low-frequency drift.

5.2 Piezoelectric Accelerometers

As discussed in Chapters 2 and 3, the *direct* piezoelectric effect leads to the generation of electric charge as a result of an applied stress and can be used to measure acceleration. This is usually achieved by measuring the resultant force from an accelerating seismic mass. Figure 5.1 shows a simplified schematic of a piezoelectric accelerometer and its main mechanical components. The piezoelectric element of an accelerometer is the active component, and Figure 5.1a shows that this particular device architecture operates in d_{33}-mode, where the force that is generated due to an acceleration is applied in a direction that is parallel to the polarisation direction of the piezoelectric material. When the accelerometer is subjected to an acceleration (a), a force (F) that is equal to ma acts on the piezoelectric element to produce a direct piezoelectric charge that is proportional to the applied force and acceleration. Figure 5.1b is an example of an accelerometer that operates in a shear d_{15}-mode, where the force due to the acceleration is perpendicular to the polarisation.

Piezoelectric Materials: Properties, Applications and Devices, First Edition.
Christopher R. Bowen and Andrew R. Plummer.

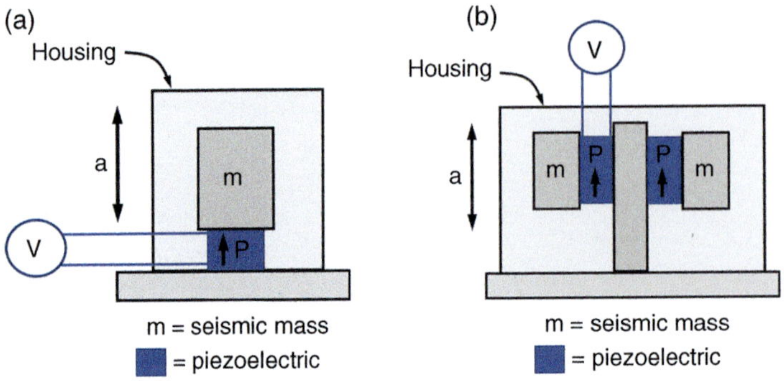

Figure 5.1 Example of a piezoelectric accelerometer. (a) A compression d_{33}-mode configuration and (b) a shear d_{15}-mode configuration. The seismic mass leads to an applied force under an acceleration that produces charge by the direct piezoelectric effect.

The *sensitivity* of the accelerometer can be defined if we consider the piezoelectric material in Figure 5.1a as a flat plate of material that can detect forces in the 3-direction.

The force (F) generated by the seismic mass (m) from an acceleration (a) is simply given by

$$F = m \cdot a \tag{5.1}$$

The piezoelectric charge (Q) generated by the force is given by

$$Q = d_{33} \cdot F = d_{33} \cdot m \cdot a \tag{5.2}$$

We can see that the charge generated is directly proportional to the acceleration. The accelerometer sensitivity can be described by the charge per unit acceleration (Q/a), for example, *pC/g*, which is

$$\frac{Q}{a} = d_{33} \cdot m \tag{5.3}$$

It can be seen, above, that the charge sensitivity is proportional to the d_{33} piezoelectric charge coefficient of the material used in the sensor and the magnitude of the seismic mass.

In addition to the measurement of charge, it is possible to measure the voltage of the sensor, such as the open-circuit voltage. In order to determine the voltage, it is first necessary to know the capacitance (C) of the piezoelectric element, which is given by

$$C = \frac{A\varepsilon_{33}^{T}\varepsilon_0}{t} \tag{5.4}$$

where A is the electrode area, t is the thickness of the piezoelectric element, ε_{33}^{T} is the relative permittivity of the piezoelectric material at constant stress in the polarisation direction (the 3-direction), and ε_0 is the permittivity of free space. Knowing the relation between charge, voltage and capacitance, the open-circuit voltage (V_o) generated by the applied force is

$$V_o = \frac{Q}{C} \tag{5.5}$$

Hence,

$$V_o = \frac{d_{33}}{\varepsilon_{33}^T \varepsilon_0} \cdot \frac{F}{A} \cdot t = \frac{d_{33}}{\varepsilon_{33}^T \varepsilon_0} \cdot \frac{ma}{A} \cdot t \tag{5.6}$$

In this case, the open-circuit voltage is proportional to the acceleration experienced by the sensor. The accelerometer sensitivity can be described by the voltage per unit acceleration, for example, V/g,

$$\frac{V}{a} = \frac{d_{33}}{\varepsilon_{33}^T \varepsilon_0} \cdot \frac{m}{A} \cdot t = g_{33} \cdot \frac{m}{A} \cdot t \tag{5.7}$$

In this case, the voltage sensitivity is proportional to the $g_{33} = d_{33}/\varepsilon_{33}^T \varepsilon_0$ piezoelectric voltage coefficient, the magnitude of the seismic mass and the thickness of the piezoelectric element. It is also inversely proportional to the area of the piezoelectric element since a small area decreases the devices capacitance to produce a higher voltage for a given charge. The fact that the d_{33} coefficient is important for the charge sensitivity of a sensor and the g_{33} coefficient is important for the voltage sensitivity of a sensor indicates that different materials may be appropriate for different modes of sensing.

We have seen that the sensitivity of an accelerometer is related to the ratio of the electrical output of the sensing element to the mechanical input. Accelerometer sensitivity is therefore often rated in terms of units of mV/g or pC/g, depending on whether the measurement signal is voltage or charge, respectively.

Note 5.1

A sensor can operate based on detecting the piezoelectric charge generated in response to a mechanical load; in this case, the piezoelectric charge coefficients (d_{ij}) are of importance. If the open-circuit voltage is used to sense the mechanical load, the piezoelectric voltage coefficients (g_{ij}) are of importance.

In terms of the frequency dependence of the sensor sensitivity, the accelerometer can be considered as a spring–mass–damper. Figure 5.2 shows that the resonant frequency of the accelerometer device, after it has been mounted, will determine the frequency range over where the accelerometer can be used reliably since, in this frequency range, it is able to maintain a constant electrical signal for a specific acceleration level. A level of consistency in terms of the sensor output is important to ensure that the sensitivity is constant over the frequency range of interest and provides *broadband* operation. Such a broadband operation is particularly important if the sensor is subjected to vibrations across a wide range of frequencies. The change in accelerometer sensitivity at resonance can be observed in Figure 5.2, which shows that the sensitivity will increase near the natural frequency of the sensor and clearly a high natural resonance frequency is therefore of interest since it will lead to a wide range of operating frequencies [1].

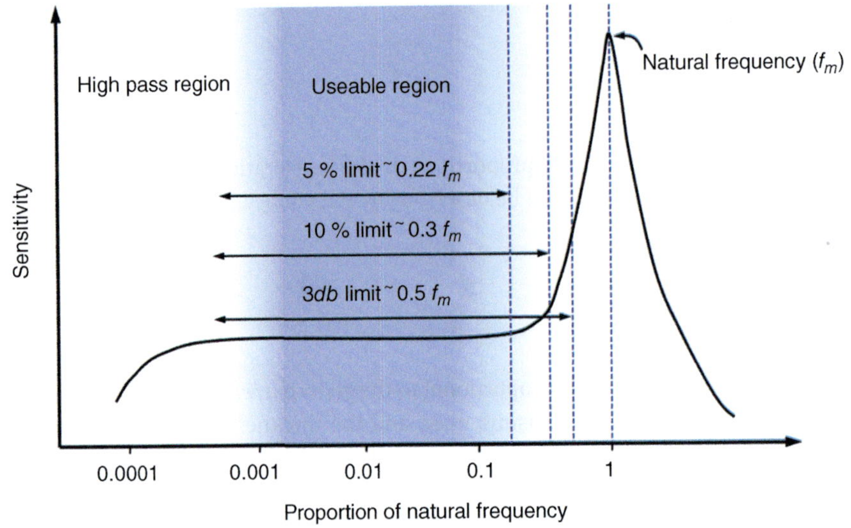

Figure 5.2 Sensitivity of the accelerometer as a proportion of its resonant frequency. Note the log-scale in frequency.

Based on the accelerometer being considered as a spring–mass–damper, a high-mounted resonance frequency corresponds to a small accelerometer device, a high-stiffness piezoelectric element or the use of a lightweight seismic mass to generate the force. Since the stiffness of many piezoelectric materials does not change significantly, the magnitude of the seismic mass is more readily changed to tune the resonant frequency. For example, while the use of a small seismic mass can increase the natural frequency, clearly the use of a low seismic mass will also lead to a reduced force being exerted on the active element, since $F = ma$ (Eq. (5.1)). As a result, the accelerometer becomes less sensitive, and therefore, there is a need to strike a balance between the sensor sensitivity and sensor bandwidth, where accelerometers with a high resonant frequency can exhibit a relatively reduced sensitivity, and high-sensitivity accelerometers exhibit a narrower frequency range of measurement.

Figure 5.2 outlines the frequency ranges and definitions of the frequency response of an accelerometer [1], which effectively indicates regions of specific tolerance. These include the following:

i) *5% frequency limit*: In this frequency limit, there is a deviation of 5% between the measured and applied vibration levels. The maximum vibration frequency that can be measured with this accuracy is approximately one-fifth ($0.22f_m$) of the mounted resonance frequency (f_m) of the accelerometer.
ii) *10% frequency limit*: In this frequency limit, there is a 10% deviation between the measured and applied vibration levels. The maximum vibration frequency that can be measured with this accuracy is approximately one-third ($0.3f_m$) of the mounted resonance frequency of the accelerometer.
iii) *3-dB frequency limit*: In this frequency limit, there is a 3-dB difference, approximately double, between the measured and applied vibration levels, where dB is defined as

$dB = 20\log_{10}(S_f/S_{ref})$, where S_f is the sensitivity of the accelerometer at a specific frequency, and S_{ref} is sensitivity at the reference frequency (e.g. 100 Hz). The maximum vibration frequency that can be measured with this accuracy is approximately one-half ($0.54f_m$) of the mounted resonance frequency of the accelerometer.

iv) *Lower frequency limit*: This frequency limit defines the lowest frequency of operation. Piezoelectric materials are best suited for dynamic loads since there are inevitably some electrical losses between the electrodes across the piezoelectric element at low frequencies due to the presence of a small, but finite, electrical conductivity in the piezoelectric material. As a result, some piezoelectric charge can be lost by electrical conduction between the electrodes. Piezoelectric sensors are therefore unable to provide a true zero-frequency response. The low-frequency limit of an accelerometer is determined by the electrical conductivity, or dielectric loss, of the material that is used and the properties of the preamplifier used to measure the electrical signal from the piezoelectric element since it determines the rate at which piezoelectric charge will leak away from the accelerometer.

Note 5.2
The natural frequency of the sensor determines the frequency range of operation, and there is a need to balance the frequency range and final sensitivity that can be achieved.

5.2.1 Accelerometer Modes of Operation (Compressive, Shear and Flexural d_{33} and d_{15})

As shown in Figures 5.1 and 5.3, a piezoelectric accelerometer can operate in various modes. In the mode where a compressive or tensile stress is applied to the piezoelectric element by a seismic mass, as shown in Figure 5.1a, the piezoelectric charge generated by the direct piezoelectric effect is collected in the polarisation direction since the device is operating in d_{33}-mode for ferroelectric ceramics (or d_{11}-mode for piezoelectric quartz) and electrodes are applied directly to the upper and lower faces of the polarised material. One of the disadvantages of this mode of operation is that any surrounding temperature fluctuations can lead to the charge being developed in the polarisation direction due to pyroelectric effects;

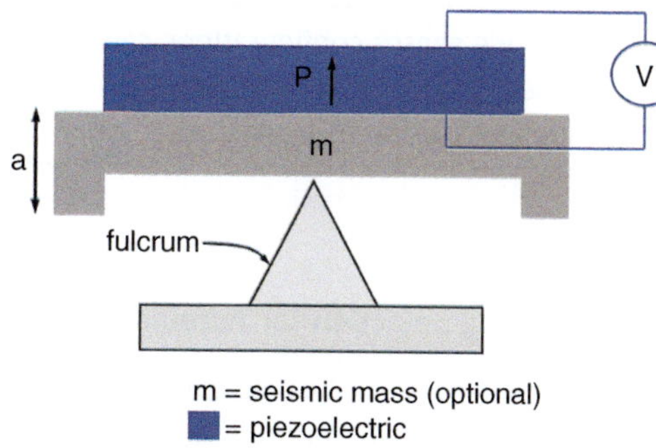

Figure 5.3 Example of a piezoelectric accelerometer in d_{31} flexure mode configuration. The seismic mass leads to an applied force under an acceleration that produces charge by the direct piezoelectric effect.

the pyroelectric coefficient (p) of a material is related to the generation of charge due to change in polarisation with temperature, rather than stress. Since a pyroelectric-induced charge can be generated at the same electrode as the piezoelectric charge in this mode, it is also collected with the vibration-induced charge and can therefore lead to reduced accuracy in the presence of thermal fluctuations. The reader can also refer to Figure 2.7, which indicates all poled ferroelectric materials are both piezoelectric and pyroelectric, while some non-ferroelectric piezoelectric materials, such as quartz, are not pyroelectric and will therefore not generate a pyroelectric charge in response to a temperature change.

In addition, in order to separate piezoelectric and pyroelectric effects, the device configuration in Figure 5.1b can be used. The use of shear deformation to operate in d_{15}-mode for ferroelectric ceramics (or d_{26}-mode for piezoelectric quartz) leads to the piezoelectric charge being collected by electrodes located on surfaces parallel to the polarisation direction. In this shear mode of operation, any pyroelectric charges that are generated on the polarised surfaces, as a result of surrounding temperature fluctuations, are not collected; this therefore reduces any thermal effects on the sensor sensitivity.

Accelerometers can also be configured to operate in a flexural/bending mode to operate in d_{31}-mode for a ferroelectric ceramic (or d_{12}-mode for piezoelectric quartz), as shown in Figure 5.3. Such a flexural configuration is naturally less stiff compared to the compression and shear mode devices in Figure 5.1, thereby leading to a reduced operating frequency. There is also an increased risk of damage due to the higher strain experienced by the piezoelectric elements, which are often brittle and of relatively low toughness. This configuration is therefore more suited for low-frequency and low-intensity vibrations, where sensor sensitivity is of primary importance.

Note 5.3

Shear mode sensor configurations apply electrodes on surfaces perpendicular to the polarisation direction. This can ensure that pyroelectric charges, which are produced on the polarised surfaces as a result of temperature fluctuations, are not collected and do not contribute to the sensor signal.

Note 5.4

Flexural mode sensor configurations can reduce device stiffness to improve sensitivity; due to their low resonant frequency and lower robustness, they are suited for low-frequency and low-intensity vibrations, where sensor sensitivity is of primary importance.

5.2.2 Material Selection for Accelerometers

A range of materials can be considered for accelerometers. One advantage of using a polycrystalline ferroelectric ceramic, such as lead zirconate titanate (PZT) or barium titanate

($BaTiO_3$), is that they can be produced in a range of desired shapes and can be subsequently poled to achieve a polarisation in a desired direction (see Chapter 2). The exact composition of the material can be varied by various doping strategies to provide specific properties (Chapter 3). We have seen in the discussion above that for charge operation (pC/g), a high d_{33} or d_{15} is desirable since the d_{ij} coefficient is a measure of the charge generation per unit force. For voltage operation (mV/g), a high g_{33} or g_{15} is desirable since g_{ij} is the electric field (and therefore voltage) per unit stress.

The Curie point (T_c) of the ferroelectric material can also be tailored by material or compositional selection to provide an upper temperature limit of operation, and effort has been made to develop high Curie temperature materials for hostile environments such as automotive, aerospace and energy and downhole for oil and gas applications; for an overview of materials with a high Curie temperature, the reader is referred to an excellent overview by Zhang et al. [2].

For non-ferroelectric and piezoelectric monocrystalline materials, such as quartz, their composition is fixed, and their shapes are more limited due to their single-crystal nature since they need to be grown and cut to the required geometry. Since these materials are not ferroelectric, the polarisation direction cannot be changed by the application of an electric field (the poling field), and therefore, the crystal must be produced in a particular orientation (or 'cuts'). Quartz crystal is an example of a non-ferroelectric piezoelectric material with a trigonal structure of class 32 symmetry [3, 4], which generates a piezoelectric charge when subject to a mechanical load. The advantage of quartz as an accelerometer material is that it has high rigidity (stiffness), thereby naturally providing a high natural frequency. Since it is not ferroelectric, the crystal is cut in particular directions relative to the polarisation axis for both pressure-sensitive and shear-sensitive elements that operate by the longitudinal effect, the transverse effect and the shear mode. The image in Figure 5.4 shows

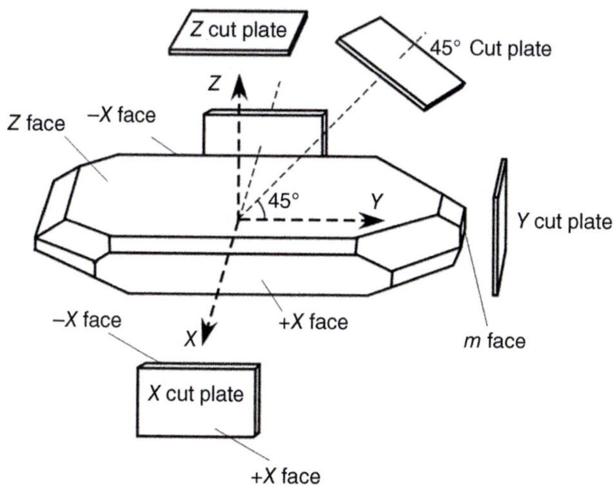

Figure 5.4 Examples of potential crystal cuts of piezoelectric quartz to create elements for operation in various compression (*X*-cut), transverse (*XY*-cut) and shear (*Y*-cut) modes. *Source:* Suzuki et al. [5] / Scientific Research Publishing Inc. / CC BY 4.0.

examples of crystal cuts and the modes of operation of the piezoelectric effect. Quartz is somewhat different from the ferroelectric piezoelectric materials, such as PZT or $BaTiO_3$, whereby in the stress-free state the crystal structure is symmetric in terms of the distribution of silicon and oxygen ions and there is initially no dipole; it is the application of a stress that leads to an asymmetry and change in polarisation. As shown in Figure 5.5a, when there is no stress applied to a quartz crystal, the electrical centre of the three silicon atoms, the electrical centre of the three oxygen atoms and the origin of the x–y plane coincide [4]. When a compressive stress is applied along the x-axis, the electrical centre of the three silicon atoms no longer coincides with the electrical centre of the three oxygen atoms. As a result, a polarisation along the x-axis is produced, as shown in Figure 5.5b. When a tensile stress is applied along the x-axis, a polarisation in the opposing direction is produced. In Figure 5.5c, a

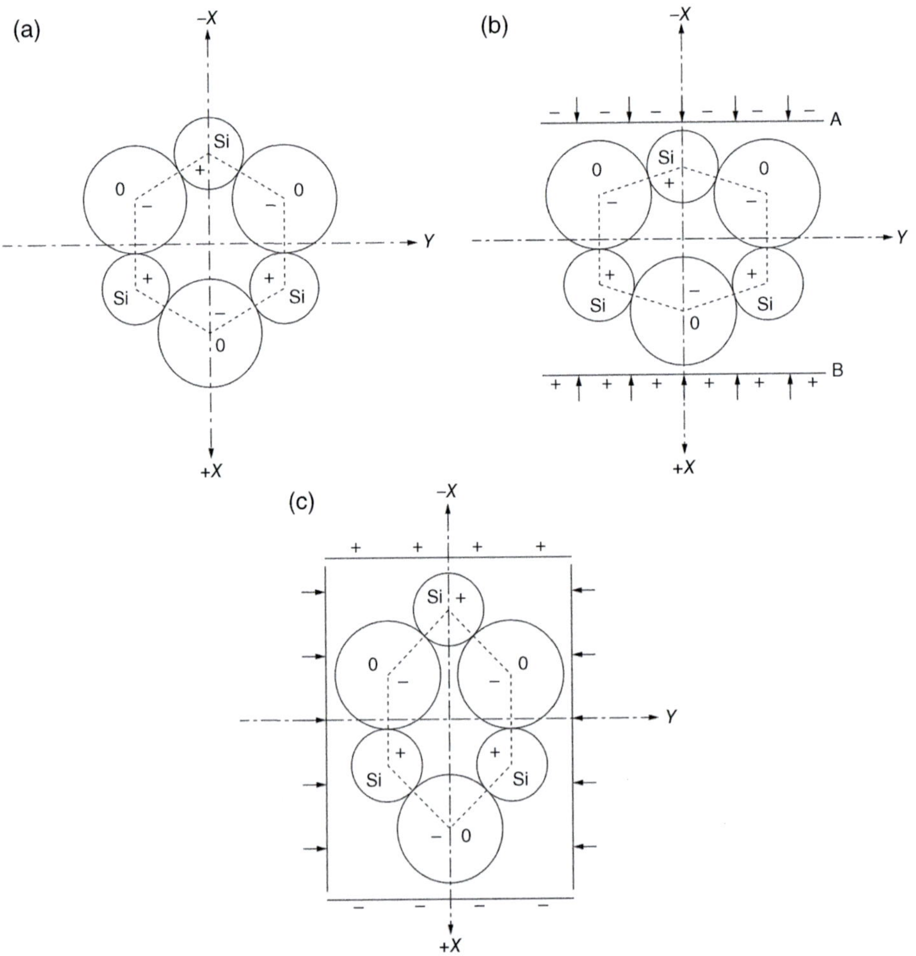

Figure 5.5 (a) Silicon and oxygen atoms in quartz viewed from the z-axis, (b) piezoelectric response due to compressive stress in the x-axis and (c) piezoelectric response due to compressive stress in the y-axis. *Source:* Saigua [4] / with permission of Elsevier.

mechanical stress is applied in the y-direction, resulting in a polarisation along the x-direction.

Quartz is of interest in terms of providing a good temperature sensitivity. In addition, since the material has no polarisation in the initial unloaded state, it is not pyroelectric (see Figure 2.7) since in the stress-free state there is no polarisation and a change in temperature therefore does not change the polarisation; this is in contrast to the poled ferroelectric materials, which are both piezoelectric and pyroelectric, since they are polarised in the stress-free state. The Curie brothers used a quartz plate that was cut perpendicular to the x-axis (called the X-cut) for their early piezoelectric experiments, and the additional information on quartz can be seen in a detailed overview by Saigua [4]. It is also of interest to note that since non-ferroelectric piezoelectric materials, such as quartz, are not poled to produce a piezoelectric response by achieving a remnant polarisation, they exhibit limited ageing compared to ferroelectric materials (discussed in Chapter 3), which naturally attempt to revert a poled ferroelectric material to its lower-energy unpolarised state.

We have seen that for high charge sensor sensitivity, there is a need for high d_{ij} charge coefficients, while for high voltage sensitivity, high g_{ij} coefficients are desirable. It is therefore of interest to compare a range of materials for piezoelectric sensing. Table 5.1 summarises the properties of some typical piezoelectric materials, including quartz, common ferroelectric ceramics and a ferroelectric polymer. It can be seen that quartz exhibits a relatively low d_{11} coefficient, especially when compared to the ferroelectric ceramics (d_{11} is used rather than d_{33} in this context due to the crystal structure of quartz). However, the g_{11} coefficient for quartz is relatively high due to its much lower permittivity compared to the ferroelectric ceramics. Quartz also has relatively high stiffness, as indicated by the low compliance in Table 5.1. Quartz is also the only non-pyroelectric material in the table, thereby limiting the potentially negative impact of temperature fluctuations on noise due to

Table 5.1 Typical properties of some common piezoelectric materials.

Parameters	Quartz	$BaTiO_3$	PZT-4 (hard)	PZT-5H (soft)	PVDF–TrFE
Type	Piezo-crystal	Ferroelectric ceramic	Ferroelectric ceramic	Ferroelectric ceramic	Ferroelectric polymer
d_{33} (pC/N)	$d_{11} = 2.3$	190	290	593	33
Relative permittivity ε_{33}^T	5	1700	1300	3400	10
g_{33} ($\times 10^{-3}$ Vm/N)	52	12.6	25.1	19.7	373
Curie temperature T_c (°C)	–	120	328	193	120
Compliance (pPa^{-1}) s_{33}^E	9.73	9.1	15.5	20.8	472
Mechanical quality Q_m	10^5–10^6	400	500	65	3–10
Pyroelectric	No	Yes	Yes	Yes	Yes

Source: [6–8].

the generation of pyroelectric charge. The ferroelectric ceramics exhibit much higher d_{33} charge coefficients compared to both quartz and the ferroelectric polymer, but their high permittivities lead to relatively low g_{33} voltage coefficients. If we compare the hard ferroelectric ceramic (which has limited domain mobility, as outlined in Chapter 3) and the soft ferroelectric ceramic (which has high domain mobility, as outlined in Chapter 3), the soft ferroelectric exhibits the highest d_{33} charge coefficients; however, the g_{33} coefficient of the soft ferroelectric is lower than the hard ferroelectric due to its high permittivity. The hard ferroelectric also has a higher Curie point (T_c) than the soft ferroelectric, making it more suitable at higher temperatures. The ferroelectric polymer, a polyvinylidene fluoride (PVDF)–trifluoroethylene (TrFE) co-polymer, exhibits a relatively low d_{33} charge coefficient but a very high g_{33} due to its low permittivity. The disadvantage of the polymer is its low stiffness (high compliance), low thermal stability and availability in limited forms, often thin films. While the polymers exhibit a low mechanical quality factor (Q_m), quartz provides a very high mechanical quality factor and limited damping.

Note 5.5

Ferroelectric ceramics typically provide high d_{ij} charge coefficients, while the lower permittivity materials such as quartz crystals and ferroelectric polymers provide high g_{ij} voltage coefficients. Ceramics and quartz provide high stiffness, with quartz providing a very high mechanical quality factor and limited damping. Quartz is non-pyroelectric, which limits the generation of additional charge due to pyroelectric effects.

5.3 Force and Pressure Sensors

In addition to acceleration, piezoelectric sensors can also be configured to measure dynamic forces and pressures. For a force sensor, a seismic mass is not required and the force acting on the sensor is applied directly to the piezoelectric material; for example, see Figure 5.6a for a d_{33}-mode force sensor. There is also potential to operate the force of pressure sensors in d_{31}- or d_{15}-mode, again to tailor sensor response to the applied load or mitigate against pyroelectric contributions by exploiting the d_{15}-mode. By combining a number of piezoelectric

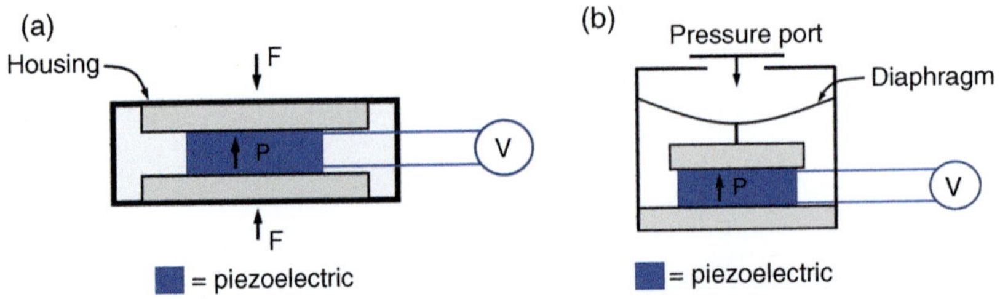

Figure 5.6 (a) Force sensor configuration. (b) Pressure sensor configuration.

elements (e.g. including extensional and shear types), a sensor device can be developed to detect three-dimensional stresses, such as those used in force plates used in biomechanics.

For a pressure sensor, a configuration is needed that allows the pressure to exert a force on the piezoelectric element from the applied pressure, for example, the use of a diaphragm as shown in Figure 5.6b. Clearly, for this case, the generated piezoelectric charge and voltage are proportional to the applied force or pressure, with the analysis of the accelerometer described above.

5.3.1 High-Frequency Capability

Applications beyond accelerometers are now described. Since piezoelectric materials develop a charge due to a change in polarisation with stress, they are particularly attractive for high-speed dynamic applications. As shown in Figure 5.2, to operate at a high frequency, there is a need for the sensor to exhibit a high natural frequency (~100 kHz), where the fabrication of small sensors with high stiffness is highly beneficial. For example, piezoelectric 'load washers' are based on high-stiffness piezoelectric quartz (see Table 5.1), which are able to measure uniaxial loads, where the sensing element is housed in a high-stiffness steel package. This high stiffness of the sensor also reduces its impact on the dynamic behaviour of the component or system being measured.

'Knock sensors' based on the piezoelectric effect are used to detect high-frequency vibrations in combustion engines. The sensors are able to continuously monitor structural vibrations in an internal combustion engine to allow adjustment of the ignition parameters during operation. In this case, the use of a flat piezoelectric knock sensor with a high resonant frequency leads to a relatively flat response in the fundamental knock frequency range, which can be in the order of 5–8 kHz [9]. There are also 'spike' sensors, where the devices resonant frequency is in the vicinity of the mean knock frequency of the engine. Such a configuration aims to exploit the high sensitivity in this region (see Figure 5.2) but at the expense of a narrower bandwidth (~10^2 Hz).

5.3.2 Sensor Sensitivity

The initial analysis described above in Section 5.2 examined the voltage and charge sensitivity for a d_{33}-mode accelerometer, where the mechanical load that is being sensed is parallel to the polarisation direction. This is the simplest case since the area subjected to the applied force (A_L), which defines the stress, is the same as the electrode area (A_E), which defines the charge density (Q/A_E) and capacitance (C); this is the case in Figure 5.7a.

We now consider sensor cases when the area subjected to the applied force and the electrode area are not equal, $A_L \neq A_E$. In Figure 5.7b, the sensor is operating in a transverse d_{31}-mode. The charge density per unit stress can therefore be described by

$$\frac{Q}{A_E} = d_{31} \cdot \frac{F}{A_L} \tag{5.8}$$

Based on the dimensions in Figure 5.7,

$$\frac{Q}{wb} = d_{31} \cdot \frac{F}{bt} \tag{5.9}$$

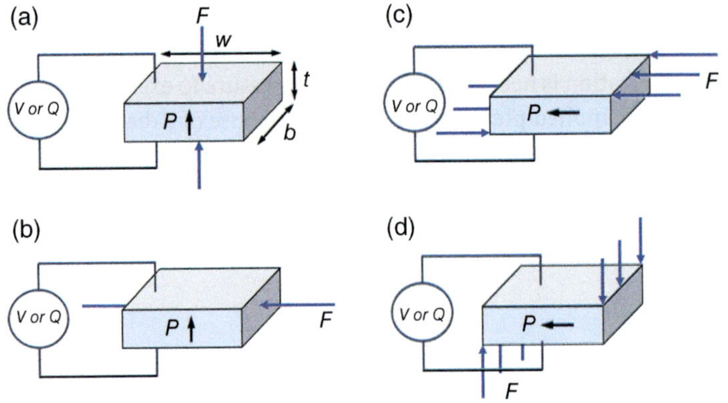

Figure 5.7 Sensor modes. (a) Parallel d_{33}-mode, (b) transverse d_{31}-mode, (c) parallel shear d_{15}-mode and (d) transverse shear d_{15}-mode.

As a result, the piezoelectric charge due to an applied force is given by

$$Q = d_{31} \cdot \frac{Fw}{t} \tag{5.10}$$

The open-circuit voltage can also be determined from the relationship $Q = CV$, where the sensor capacitance is

$$C = \frac{A_E \varepsilon_{33}^T \varepsilon_0}{t} = \frac{wb\varepsilon_{33}^T \varepsilon_0}{t} \tag{5.11}$$

As a result, the open-circuit voltage (Vo) is given by

$$V_o = \frac{d_{31}}{\varepsilon_{33}^T \varepsilon_0} \cdot \frac{F}{b} = g_{31} \cdot \frac{F}{b} \tag{5.12}$$

Interestingly, it can be seen that $Q \propto w/t$, while $V \propto 1/b$, indicating that the geometry of the piezoelectric element can be used to tailor the sensor sensitivity.

A similar analysis can be undertaken to examine a sensor operating in parallel shear d_{15}-mode, as shown in Figure 5.7c. The charge density per unit stress can be described by

$$\frac{Q}{A_E} = d_{15} \cdot \frac{F}{A_L} \tag{5.13}$$

Since in this case $A_E = A_L$, the charge is simply given by

$$Q = d_{15} \cdot F \tag{5.14}$$

In this case, the electrodes are perpendicular to the poling direction and the capacitance is now related to the relative permittivity at constant stress normal to the polarisation direction, ε_{11}^T, where

$$C = \frac{A_E \varepsilon_{11}^T \varepsilon_0}{t} = \frac{wb\varepsilon_{11}^T \varepsilon_0}{t} \tag{5.15}$$

The open-circuit voltage (Vo) is therefore given by

$$V_o = \frac{d_{15}}{\varepsilon_{11}^T \varepsilon_0} \cdot F \frac{t}{wb} \tag{5.16}$$

So that,

$$V_o = g_{15} \cdot F \cdot \frac{t}{wb} \tag{5.17}$$

Finally, in a transverse shear d_{15}-mode, the charge density per unit stress can be described by

$$\frac{Q}{A_E} = d_{15} \cdot \frac{F}{A_L} \tag{5.18}$$

$$\frac{Q}{bw} = d_{15} \cdot \frac{F}{bt} \tag{5.19}$$

$$Q = d_{15} \cdot F \cdot \frac{w}{t} \tag{5.20}$$

The open-circuit voltage is, therefore,

$$V_o = \frac{d_{15}}{\varepsilon_{11}^T \varepsilon_0} \cdot \frac{F}{b} = g_{15} \cdot \frac{F}{b} \tag{5.21}$$

Note 5.6

Examination of the charge and voltage generated under different sensor configurations indicates the important role of materials selection, where high d_{33}, d_{31} or d_{15} charge coefficients are needed for high-charge sensor sensitivity, while high g_{33}, g_{31} or g_{15} voltage coefficients are needed for high-voltage sensor sensitivity. The geometry of the piezoelectric element is also important, where the dimensions can influence the generated charge or voltage, depending on device configuration.

5.4 Temperature and Thermal Effects

The sensitivity of an accelerometer or sensor can change with operating temperature since there can be changes in the mechanical properties, such as stiffness, and the electrical properties, such as permittivity and dielectric loss, and the d_{ij} and g_{ij} coefficients can change. Figure 5.8 shows an example of the variation in capacitance (related to material permittivity), charge sensitivity (which is related to the piezoelectric d_{ij} coefficients) and voltage sensitivity (which is related to the piezoelectric g_{ij} coefficients, $g_{33} = d_{33}/\varepsilon_{33}^T\varepsilon_0$) of a typical ferroelectric and piezoelectric material. The observed change in sensitivity is well defined, and if the temperature of operation is sufficiently below the Curie point of a ferroelectric material (T_c), no permanent change occurs in sensor performance. However, if the

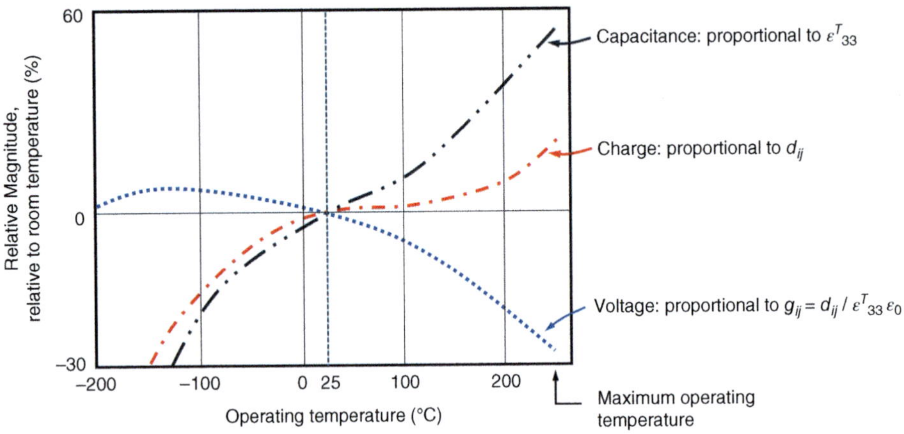

Figure 5.8 Temperature effects of capacitance (related to permittivity), charge (related to d_{ij}) and voltage sensitivity j (related to g_{ij}; $g_{ij} = d_{33}/\varepsilon_{33}^T\varepsilon_0$). *Source:* Adapted from Serridge and Licht [1].

accelerometer is subjected to a temperature close to its T_c, there can be a permanent change in the remnant polarisation due to depolarisation (Chapter 3) and therefore sensitivity. As a result, each accelerometer will have a specified maximum operating temperature; at a higher temperature, the piezoelectric element will begin to depolarise due to a loss of polarisation as a result of domain switching due to a combination of temperature and stress, leading to a permanent change in sensitivity. It is of interest to note in Figure 5.8 that as the temperature increases, the capacitance will increase as a result of an increase in permittivity (ε_{33}^T) since a high temperature facilitates an easier polarisation under an applied electric field. The charge is related to the d_{ij} coefficient, and it is less sensitive to temperature but tends to increase with temperature since the higher temperature leads to a greater change in polarisation with stress and therefore a higher d_{ij} charge coefficient. The voltage, which is related to the g_{ij} coefficient, decreases with increasing temperature and mirrors the temperature dependence of capacitance due to the relationship $g_{ij} = d_{ij}/\varepsilon_{33}^T\varepsilon_0$.

In addition to the sensitivity of an accelerometer changing with operating temperature, fluctuations in the ambient temperature, such as any airflow during a measurement, can lead to the generation of low-frequency noise. As stated in Section 5.2.1, the pyroelectric effect will also generate charge as a result of a change in the level of polarisation of a material due to a temperature change. This effect is of particular concern in poled ferroelectric materials, such as the PZT ceramics or other poled ferroelectrics, which are both piezoelectric and pyroelectric (see Table 5.1 and Figure 2.7). Since these pyroelectric charges are developed on the surfaces of the material perpendicular to the poling direction, as in d_{33}-mode sensors, it is less of a concern for shear mode (d_{15}) sensors, where the electrodes are parallel to the poling direction.

Thermal expansion during temperature changes can also lead to changes in sensor sensitivity, where non-uniform thermal expansion can lead to sections of the accelerometer deforming at different rates, leading to the development of stress in the piezoelectric element, leading to the generation of a piezoelectric charge. The thermal expansion of a piezoelectric has been discussed to some extent in Chapter 3 (Sections 3.12 and 3.13). Clearly, thermal expansion effects can occur in all piezoelectric materials, even in non-pyroelectric materials, such as quartz.

5.5 Hydrophones

5.5.1 Background to Hydrostatic Coefficients

We have seen that piezoelectric materials are a class of active materials that are employed in a wide range of applications, including accelerometers and force and pressure sensor applications. When used as a sensing element, piezoelectric voltage constants, which relate to the electric field generated per unit stress and include g_{33}, g_{31} and g_{15}, are important material properties to determine the voltage sensitivity of the sensor device. If the charge is being used as a measurement signal, then the piezoelectric charge coefficients such as d_{33}, d_{31} and d_{15} are more relevant. To now, we have considered a uniaxial load on the piezoelectric sensing element, although the direction has been varied. However, we will now consider a specific application, whereby the mechanical load is hydrostatic in nature and can be applied to multiple faces of the piezoelectric element.

Hydrophones are underwater receivers and can operate under hydrostatic conditions, especially when attempting to detect low-frequency (large wavelength) signals. These are an important type of piezoelectric transducers that are used to detect acoustic signals underwater and use the direct piezoelectric effect to convert mechanical loads from acoustic waves into some form of electrical signal, such as charge or voltage.

When a piezoelectric material is being used in a hydrophone, key material properties include [10]:

i) the hydrostatic charge (d_h) coefficient, which can be related to the actuation (transmit) capability of the piezoelectric material,
ii) the hydrostatic voltage (g_h) coefficient, which can be related to the sensitivity (V/Pa) of a hydrophone, and
iii) the hydrostatic figure of merit ($FoM_1 = d_h \cdot g_h$), which can be related to the signal-to-noise ratio of a hydrophone.

When a piezoelectric device is operating at frequencies well below the resonance frequency (resonance has been outlined in Chapter 3), energy dissipation due to dielectric loss ($\tan \delta$) is also important and an additional FoM, defined by $FoM_2 = (d_h \cdot g_h)/\tan \delta$, can be used.

We will see in this chapter that the hydrostatic FoMs ($FoM_1 = d_h \cdot g_h$ or $FoM_2 = (d_h \cdot g_h)/\tan \delta$) can be determined by knowing the piezoelectric charge coefficients (d_{ij}) and the permittivity of the material, due to the relationships $d_h = d_{33} + 2d_{31}$ and $g_h = d_h / \varepsilon_{33}^T \varepsilon_0$, where d_{33} and d_{31} are the longitudinal and transverse piezoelectric charge coefficients, ε_{33}^T is the relative permittivity at constant stress, and ε_0 is the permittivity of the free space. These relatively simple relationships indicate that key requirements to achieve a high hydrostatic performance are to select, or even design, a piezoelectric material or structure that has both a hydrostatic charge coefficient (d_h) and a low relative permittivity at constant stress (ε_{33}^T).

For most, but not all, ferroelectric ceramics, it can be assumed that $d_{33} \approx -2d_{31}$. This results in many piezoelectric ceramics exhibiting a low d_h hydrostatic charge coefficient; in addition, piezoelectric ceramics also tend to have a high relative permittivity. As a result, many monolithic piezoelectric ceramic materials have a low d_h charge coefficient, low g_h voltage coefficient and a low $d_h \cdot g_h$ FoM, which limits performance for low-frequency hydrophones subject to hydrostatic loads.

5.5.2 Derivation of Performance Indicators for Hydrophone Materials

There are various FoMs that describe the performance of a piezoelectric material for specific hydrophone applications; the following section outlines in more detail these performance FoMs, which will be defined for specific applications, in particular those that are related to operation under hydrostatic conditions.

Figure 5.9 provides a visual indication of how the wavelength of an acoustic signal is related to its frequency when in water; here, the speed of sound in water is assumed to be 1500 m/s. Hydrophones are often attempting to detect low-frequency signals, for example 10 kHz and smaller, since they travel the furthest distance in the water due to the lower attenuation at low frequencies. At frequencies below 10 kHz, the wavelengths in Figure 5.9 are over 15 cm and are therefore larger than the piezoelectric material used in a hydrophone device or the device itself. At frequencies of 1 kHz, the wavelength is in excess of 1 m; see Figure 5.9. Since the wavelength is larger than the device, we can assume that the mechanical load on the piezoelectric element at low frequency is hydrostatic in nature, as the device is sitting within individual compressions or rarefactions of the low-frequency acoustic wave.

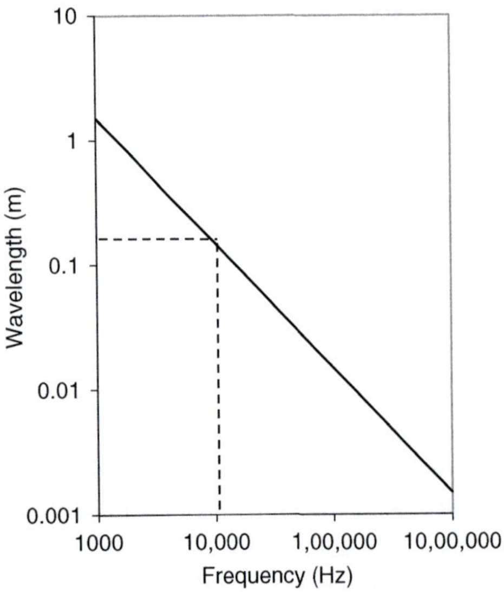

Figure 5.9 Acoustic wave frequency versus wavelength in water showing relatively large wavelengths at low frequencies; in this figure, it is assumed that the speed of sound in water is 1500 m/s. At 1 kHz, the wavelength is over 1 m and is over 0.1 m for 10 kHz. *Source:* Perry [10] / with permission of Andrew J Perry.

For the piezoelectric effects to be quantified under hydrostatic conditions, the relevant coefficients should be understood, where d_h is the *hydrostatic strain coefficient* (relevant if you are considering the material as a transmitter of acoustic signals) or the *hydrostatic charge constant* (relevant if you are considering the materials as a charge sensor). In a similar way to the accelerometers and pressure/force sensors described above, the voltage can also be used to sense a mechanical load, and therefore, the g_h, which is termed the *hydrostatic voltage constant*, is also of relevance.

To overview the response of a piezoelectric under hydrostatic conditions, it is of interest to initially consider the response of the material when it is used as an actuator in terms of hydrostatic performance, where the applied electric field and the poling direction are indicated in Figure 5.10a. The piezoelectric coefficient d_{ij} can be defined as the strain in the j direction, divided by the electric field that has been applied in the i direction. As a result of the applied electric field, the piezoelectric will exhibit a positive strain in the 3-direction (the polarisation direction, z), which is defined by the d_{33} coefficient. The electric field with the 3-direction will also lead to a negative strain in the 1-direction (x-direction) and the 2-direction (y-direction), which is related to the d_{31} coefficient (since $d_{31} = d_{32}$ in most ferroelectric ceramics). In essence, as the material extends in the 3 (or z)-direction, which is the poling direction, it will also contract in the 1 (or x)-direction and 2 (or y)-direction.

As a result of these strains in the 3-, 2- and 1-directions, the net hydrostatic strain is the sum of the strains in the three directions. For many, but not all, ferroelectric ceramic materials, the magnitude of the d_{31} coefficient is typically half that of the d_{33} coefficient, so that $d_{33} \approx -2d_{31}$. As a result, there is a *small hydrostatic strain* and a small overall change in the volume of the piezoelectric due to the electric field. For instance, for PZT-5H, a soft ferroelectric PZT, the d_{33} is ~590 pm/V (or pC/N), while the d_{31} is ~270 pm/V (or pC/N); this leads to a small d_h ($= d_{33} + 2d_{31}$) of only ~50 pm/V (or pC/N).

By considering the material as an actuator, as shown in Figure 5.10a, the low hydrostatic strain therefore originates from the positive piezoelectric strain in the 3-direction

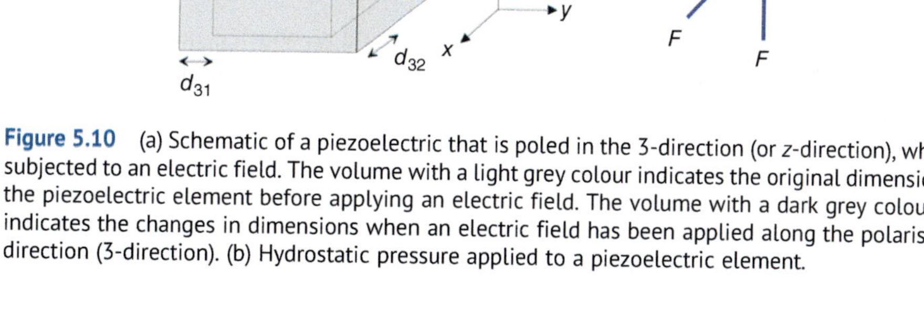

Figure 5.10 (a) Schematic of a piezoelectric that is poled in the 3-direction (or z-direction), which is subjected to an electric field. The volume with a light grey colour indicates the original dimensions of the piezoelectric element before applying an electric field. The volume with a dark grey colour indicates the changes in dimensions when an electric field has been applied along the polarisation direction (3-direction). (b) Hydrostatic pressure applied to a piezoelectric element.

(the poling direction, z) and the relatively large and negative strains in the corresponding 1- and 2-directions, which effectively cancel out; there is effectively a small overall volume change in response to an electric field.

If we now consider the material to operate as a sensor by the *direct* piezoelectric effect, a charge will be generated due to an applied hydrostatic stress, where the coefficient d_{ij} is the charge density (C/m^2) in direction j per unit applied stress (N/m^2) when it is applied in direction i. This is shown in Figure 5.10b. Therefore, the d_{33} piezoelectric charge coefficient is related to the charge that is generated per unit force in the 3-direction (poling direction) and the d_{31} piezoelectric coefficient relates to the charge developed per unit force in the 1-direction and the 2-direction. As stated, piezoelectric d_{33} and d_{31} coefficients tend to be opposite in sign since the application of a compressive stress along the poling direction (d_{33}) will lead to a reduction in the length of the crystallographic unit cell in the 3-direction/poling direction (Figure 2.2a); this will reduce the polarisation of the material. In contrast, the application of a compressive stress perpendicular to the unit cell in the 1-direction or 2-direction will increase the length of the crystallographic unit cell in the poling direction (d_{31}); this will *increase* the polarisation of the material [10].

As we are considering hydrophone sensors that are sensing low-frequency signals of large wavelengths, the behaviour of the material under hydrostatic conditions needs to be considered. To determine the hydrostatic strain that is developed per applied unit electric field (by the converse piezoelectric effect) or the charge that is generated per unit of hydrostatic force (by the direct piezoelectric effect), there is a need to determine the hydrostatic d_h coefficient, which is defined by

$$d_h = d_{33} + d_{32} + d_{31} \tag{5.22}$$

Figure 5.6 shows that d_{31} and d_{32} are both opposite in sign compared to the d_{33} coefficient; likewise, a compressive force in the 3-direction (z-direction) will produce an electric field in an opposite direction to the field generated by the compressive force in the 1-direction and the 2-direction. Many, but not all, of the well-known ferroelectric materials are transversely isotropic, and $d_{31} = d_{32}$ and d_h can be simplified to

$$d_h = d_{33} + 2d_{31} \tag{5.23}$$

The d_h coefficient has units of charge per unit force (C/N) when employed as a hydrophone. Hydrophone sensitivity is often defined in volts per Pascal (V/Pa); as a result, the g_h hydrostatic voltage coefficient (where $g_h = d_h/\varepsilon_{33}^T\varepsilon_0$) is better suited for considering the performance of a materials sensing. The g_h coefficient is related to the electric field that is generated per unit of hydrostatic stress (V/m Pa).

It is of interest to consider the relationship between d_h and g_h, which is now derived. The d_h represents the charge generated by the direct piezoelectric effect per unit applied hydrostatic force; therefore,

$$d_h = \left(\frac{Q}{F_h}\right) \tag{5.24}$$

Here, Q is the charge (C) generated, and F_h represents the hydrostatic force (N).

If we consider the piezoelectric element as a simple flat plate and the hydrophone sensor as a parallel plate capacitor, where the capacitance of the piezoelectric can be described by

$$C = \frac{A\varepsilon_{33}^{T}\varepsilon_0}{t} \tag{5.25}$$

where t is the thickness (m), A is the upper/lower surface area (m^2), ε_{33}^{T} is the relative permittivity at constant stress along the direction of polarisation (3-direction), and ε_0 is the permittivity of free space (8.85×10^{-12} F/m). From the relationship $Q = CV$, the open-circuit voltage (Vo) generated as a result of a piezoelectric charge, Q, from a hydrostatic load is given by

$$V_o = \frac{Qt}{A\varepsilon_{33}^{T}\varepsilon_0} \tag{5.26}$$

By combining Eqs. (5.3) and 5.5), the voltage can be described by

$$V_o = \frac{d_h}{\varepsilon_{33}^{T}\varepsilon_0} \cdot \frac{F}{A} \cdot t \tag{5.27}$$

The equation can be rearranged to determine the hydrophone receive sensitivity (V/σ), which is the voltage developed per unit acoustic pressure/stress, σ (V/Pa). Since pressure $\sigma = F/A$, then

$$\frac{V_o}{\sigma} = \frac{d_h}{\varepsilon_{33}^{T}\varepsilon_0} \cdot t \tag{5.28}$$

For a hydrophone of a specific thickness, the hydrophone sensitivity is therefore proportional to the material properties of $g_h = d_h/\varepsilon_0\varepsilon_{33}^{T}$. The hydrostatic voltage constant (g_h) can be described as the electric field (V/t) that is generated per unit of hydrostatic stress (F/A). The linkage between g_h, d_h and ε_{33}^{T} can be determined from Eq. (5.6) to give

$$g_h = \left[\frac{V}{t}\right]\left[\frac{A}{F}\right] = \frac{d_h}{\varepsilon_{33}^{T}\varepsilon_0} \tag{5.29}$$

As a result of the simple analysis, to produce a *high-sensitivity hydrophone* (that generates a large voltage per unit of hydrostatic stress), the material will ideally develop a large amount of charge per unit of hydrostatic force (d_h) while also having a low relative permittivity at constant stress (ε_{33}^{T}); this combination of properties will provide a high g_h. We will see that materials designs and piezoelectric composite designs are often undertaken to produce new materials with a combination of both a high d_h and a low ε_{33}^{T}.

Based on our knowledge that a low d_h is a result of the small volume change in many piezoelectric materials, since $d_{33} \approx -2d_{31}$, it is intuitive to consider that high-d_h piezoelectric materials are likely to have d_{31} coefficients of low magnitude.

The value of the hydrostatic voltage coefficient, g_h, can therefore be used as a performance indicator to allow materials to be evaluated, selected and even designed for high-sensitivity hydrophones. In addition to selecting bulk materials with high g_h, the need for a high d_h and low ε_{33}^{T} is the basis of creating a range of *piezocomposite* structures, which is outlined later in this chapter.

From Eq. (5.29), the receive sensitivity ($M=V/\sigma$) of a hydrophone device operating under hydrostatic conditions is related to g_h and the thickness (t) of the piezoelectric element,

$$M = g_h{}^{*}t \ \left(\mathrm{V\,Pa^{-1}}\right) \tag{5.30}$$

This can also be defined as

$$M = -20\log(g_h{}^{*}t) \ \left(\mathrm{dB\ re\ 1V\,\mu Pa^{-1}}\right) \tag{5.31}$$

In addition to the hydrophone sensitivity, it is also of interest to minimise noise and maximise the signal-to-noise ratio. In the low-frequency approximation, the piezoelectric hydrophone is considered to be small in relation to the acoustic wavelength that is being measured so that the hydrophone can be assumed to be operating at frequencies well below mechanical resonance. In this case, the signal-to-noise ratio can be minimised by having a larger volume of material, where a detailed analysis has been presented by Young [10]. However, for a fixed volume of materials, this can be achieved by selecting materials with a d_h and g_h as large as possible [11]. This leads to the *hydrophone figure of merit (FoM_h)*; the derivation is also described in detail in a textbook by Wilson [3].

$$FoM_h = d_h \cdot g_h = d_h^2/\varepsilon_{33}^T\varepsilon_0 \tag{5.32}$$

Interestingly, in the next chapter (Chapter 6) on energy harvesting, the product $d_{ij}.g_{ij}$ is used to determine the electrical energy per unit stress, thereby indicating a link between signal-to-noise ratio and electrical energy generation.

The dielectric loss (tan δ) is also a factor since a high tan δ can lead to a loss of charge between the electrodes on the piezoelectric surfaces. As a result, the parameter $d_h{\cdot}g_h/\tan\delta$ is also considered as a relevant FoM, which clearly indicates that a low dielectric loss is beneficial to reducing noise. Noise is often a limiting factor for hydrophones; therefore, the ratio of signal energy to noise energy is an important factor. A wide variety of configurations are potential for hydrophone design, and the reader is referred to excellent texts in this area [3, 11, 12].

Note 5.7

For low-frequency hydrophone applications, hydrostatic parameters are important, depending on the need to measure charge (d_h) or voltage (g_h). A high-$d_h.g_h$ FoM is also advantageous in order to attain a high signal-to-noise ratio for a specific volume of material.

5.5.3 Hydrophone Construction

A piezoelectric hydrophone can be considered to be an underwater microphone. Acoustic waves in water lead to localised regions of high pressure and low pressure, which are directed onto the piezoelectric material in a hydrophone. As a result of the direct piezoelectric effect, the pressure variations are converted into an electrical signal. This piezoelectric

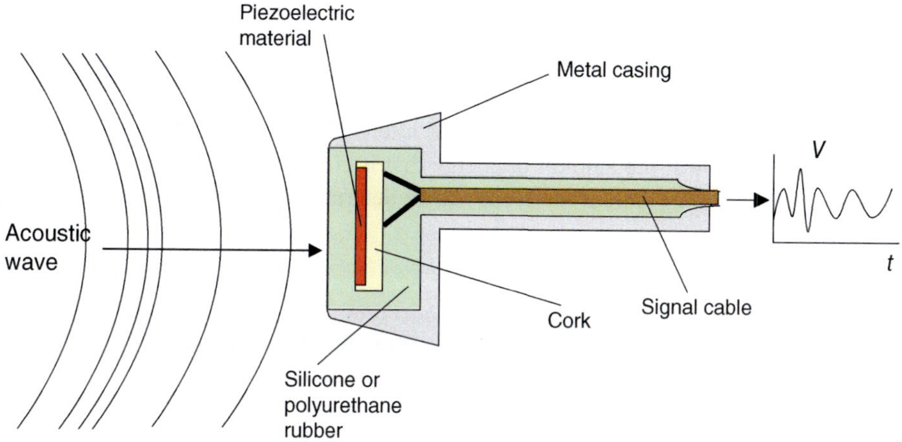

Figure 5.11 Diagram of a piezoelectric hydrophone, where acoustic waves (on the left) use the direct piezoelectric effect to produce an electrical signal (on the right). *Source:* Perry [10] / with permission of Andrew J Perry.

signal (charge or voltage) is then passed to a preamplifier and analysed; Chapter 7 outlines the use of signal conditioning for a piezoelectric sensor; see Section 7.7.

Figure 5.11 is a simple diagram of a conventional hydrophone device. When placed in water, an acoustic signal initially interacts with an acoustic matching interlayer which is typically manufactured from a silicone polymer or other rubbery material; the acoustic pressure is then passed on to the piezoelectric sensing element. Since a high sensitivity is required, which is often defined with units of V/Pa, it is often beneficial to produce a large field per unit of hydrostatic stress; this is related to the hydrostatic g_h voltage coefficient of the piezoelectric material.

For a hydrophone device to operate successfully in water, there is a need for efficient transfer of the acoustic signal into the piezoelectric sensing element, where this can be optimised by judicious use of an *impedance matching* layer [13]. The difference in acoustic impedance between the two materials or media of interest, such as the piezoelectric ceramic and water, will have a direct impact on how much energy is transferred into the piezoelectric material or is reflected by the hydrophone–water interface and lost. The acoustic impedance of a material is defined as the product of the density and the sound velocity (V_s),

$$Z = \rho V_s \tag{5.33}$$

Z represents the acoustic impedance of the material/fluid and is often measured in units of 'rayl', ρ is material/fluid density, and V_s is the velocity of sound of the material/fluid.

Eq. (5.13) provides the relationship between the velocity of sound of a material, its elastic properties and its density,

$$V_s = \left(\frac{E}{\rho}\right)^{1/2} \tag{5.34}$$

where E is the elastic modulus. Substituting Eq. (5.12) into Eq. (5.13) leads to:

$$Z = (\rho E)^{1/2} = \rho V_s \tag{5.35}$$

If the acoustic impedance of each medium of interest (ceramic, water or impedance matching layer) is known, the amount of energy that is either reflected or transmitted at an interface between two different materials can be determined via

$$R = \left(\frac{Z_2 - Z_1}{Z_2 + Z_1}\right)^2 \tag{5.36}$$

where R is the fraction of energy that is reflected off the interface between the two media, and Z_1 and Z_2 represent the acoustic impedances of the two media of interest: for example (i) ceramic–water, (ii) matching layer–water and (iii) matching layer–ceramic.

It is now possible to use Eqs. (5.33)–(5.36) to make some estimations of R based on the properties of water ($\rho \sim 1000\ \text{kg/m}^3$, $V_s \sim 1500\ \text{m/s}$, $Z \sim 1.5 \times 10^6\ \text{kg/m}^2\text{/s}$ or 1.5 Mrayl) and a soft PZT ceramic, such as PZT-5H (E ~ 60 GPa and $\rho \sim 7500\ \text{kg/m}^3$, $Z \sim 21 \times 10^6\ \text{kg/m}^2\text{/s}$ or 21 Mrayl) [10]. Based on these properties, for a piezoelectric ceramic–water interface as shown in Figure 5.12, approximately 75% of the acoustic energy is reflected. The large amount of reflection (R ~ 75%) is a result of the large differences in the acoustic impedance between the ceramic sensing element and water, namely a large ($Z_1 - Z_2$); these two materials can therefore be considered to have a low level of impedance matching. Clearly, good energy transfer between two different media and a low reflection (R) can be achieved by minimising the difference in acoustic impedances between the two materials/media, namely a low ($Z_1 - Z_2$); see Eq. (5.36).

To improve the degree of energy transfer between the piezoelectric ceramic and water, an additional material that exhibits an acoustic impedance that is somewhere between the two materials/media of interest can be applied to the ceramic–water interface. For

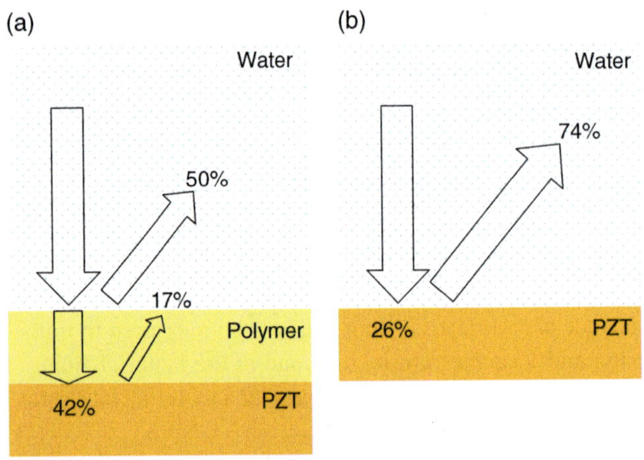

Figure 5.12 Diagram demonstrating reflection of acoustic energy due to acoustic mismatch between materials (a) with and (b) without an impedance matching layer. *Source:* Perry [10] / with permission of Andrew J Perry.

ceramic–water interfaces, potential impedance-matching-layer materials include a silicone elastomer or a polymer that has both a low stiffness and density. For a polymer with a density of 3000 kg/m^3 and an elastic modulus of $E = 20$ GPa, the acoustic impedance (Z) is approximately ~7.7×10^6 $kg/m^2/s$. This fraction of acoustic energy that is transmitted (not reflected) at the water–polymer interface is $R \sim 50\%$, and subsequently, the transmitted energy at the polymer–ceramic interface is $R \sim 83\%$. As a result, 42% of energy is transmitted from water to the ceramic piezoelectric sensing element. This impact of introducing an impedance matching interlayer between the ceramic (PZT) sensor and water is shown in Figure 5.12. Impedance matching layers are also employed for applications related to structural health monitoring and medical transducers, where there are clearly differences in the acoustic impedance between biological tissues with a low stiffness and density and the high stiffness and high density of the ceramic element [10].

Note 5.8

The high stiffness, density and velocity of sound of ceramics often mean they are poorly matched with water or biological tissue. Impedance matching layers can reduce the impedance mismatch between the sensing element and the media it is sensing from.

Clearly from the analysis if the elastic properties and density of a piezoelectric can be reduced to a value that is closer to that of water, a greater amount of acoustic energy would be transmitted between the water and ceramic; as a result, the fraction of transmitted energy could be increased or the need for a matching layer could be avoided. One approach to achieve this goal is to combine a piezoelectric ceramic with a polymer phase to create a piezoelectric composite, often termed a *piezocomposite*. The range of potential composite structures and their properties are now examined.

5.6 Piezocomposite Sensors

Figures of merit and material properties for the selection of monolithic piezoelectric materials have been discussed for applications related to both sensing and hydrophones. For hydrophones, the impact of d_h, g_h, $d_h{\cdot}g_h$ and the acoustic impedance (Z) has been outlined. Rather than selecting monolithic piezoelectric materials from the range of available compositions outlined in Chapter 3 (e.g. see Navy type compositions outlined in Table 3.11), another route to produce new materials is to tailor the properties by making the piezoceramic a component of a composite structure and create a 'piezocomposite'. The other components of the piezocomposite can be active, such as other piezoelectric materials, or passive, such as a polymer or even air. In a similar way to structural composites, such as carbon fibre-reinforced composites, the structure and arrangement of each phase in the composite are important in determining the composite behaviour. For a piezocomposite, the nature of the interconnections between the two (or more) phases in the composite is of particular importance, typically to maintain a high level of polarisation.

To date, we have discussed piezoelectric devices with a focus on producing or selecting monolithic piezoelectric elements of high density. Some piezoelectric materials of low density, such as highly porous piezo-foams or piezoelectric ceramic–polymer mixes, can exhibit large d_h and g_h coefficients when subject to hydrostatic loads. For a simple two-phase system (e.g. a ceramic and polymer mix), there are various configurations in which the phases can be connected, and these contrasting arrangements can be defined by describing the degree of 'connectivity' of the phases. The nomenclature used to describe connectivity patterns is now described and was first introduced by the pioneering work of Newnham et al. [14].

Each individual component, or phase, in the piezoelectric composite will have zero-, one-, two- or three-dimensional connectivity to itself. For a two-phase system, this connectivity is therefore defined by two numbers, for example, '*X*–*Y*'. Examples of potential connectivity patterns for a two-phase piezocomposite are outlined in Figure 5.13. The first number (*X*) is often related to the piezoelectric phase in the composite, and the second number (*Y*) is related to the passive phase, which can be a polymeric phase with low stiffness, low density and low permittivity. A good example is a '1–3' piezocomposite which consists of rods, or pillars, of a piezoelectric ceramic that is connected in one dimension and that is embedded in a continuous polymer phase which is connected in three dimensions. Likewise, a '0–3' piezocomposite consists of individual and isolated piezoelectric spheres, powders or particulates that are embedded in a continuous and three-dimensionally connected polymer phase; see Figure 5.14.

There are many ways to produce a piezocomposite. Figure 5.14 shows some of the most common configurations of composites along with their respective connectivity. In each case, the properties of piezocomposite depend on the architecture, volume fraction and properties of each phase [10].

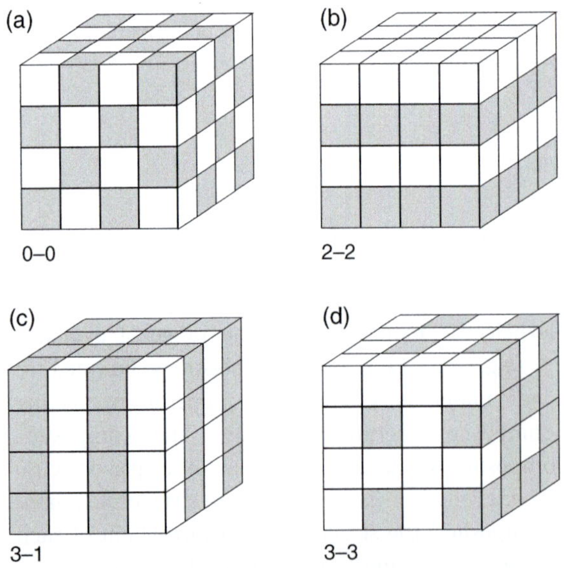

Figure 5.13 Schematic of piezoelectric composite structures with different types of connectivity. *Source:* Newnham et al. [14] / with permission of Elsevier.

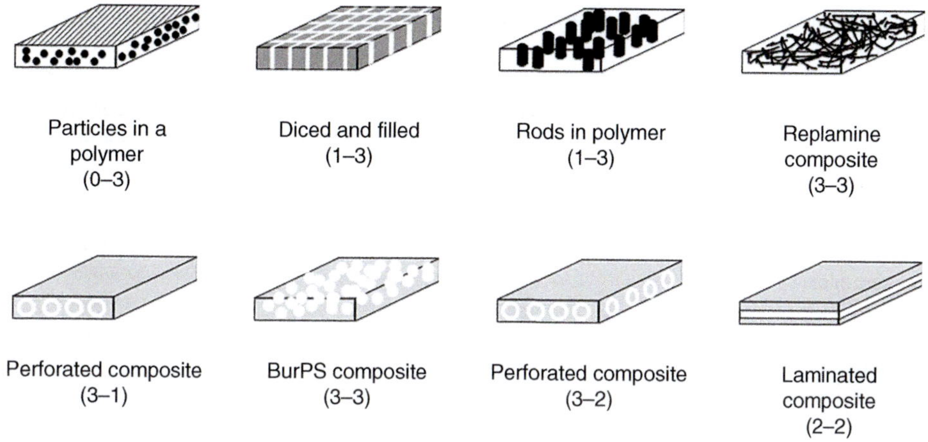

Figure 5.14 Examples of piezoelectric composites and their corresponding connectivities. *Source:* Perry [10] / with permission of Andrew J Perry.

We have outlined that to maximise the sensitivity of a low-frequency hydrophone, the hydrostatic voltage coefficient, g_h, should be as large as possible. An increase in the g_h voltage coefficient can be achieved by producing a composite material with a low permittivity, ε_{33}^T, or producing a high-d_h composite structure (see Eq. (5.29)), or ideally both. One advantage of piezoelectric composites is that if a polymer phase is used to form a composite, it will have a significantly lower permittivity compared to the piezoelectric ceramic phase; the use of a polymer will therefore lower the effective (overall) composite permittivity and increase the g_h coefficient. The d_h hydrostatic charge coefficient can also be increased by reducing the magnitude of the d_{31} coefficient *relative* to the magnitude of the d_{33} coefficient by creating a composite structure that effectively *decouples* the d_{33} and d_{31} coefficients. For example, in a '1–3' composite, the effective d_{33} coefficient does not decrease significantly when a dense piezoelectric ceramic is replaced by a series of piezoelectric rods, while the magnitude of the d_{31} coefficient decreases significantly; this leads to a high d_h, where the combination of a high d_h coefficient and a low permittivity can lead to very large increases in g_h (voltage, Eq. (5.29)) and $d_h.g_h$ (signal-to-noise ratio, Eq. (5.32)).

Table 5.2 summarises and compares a range of ferroelectric ceramics, polymers and piezoelectric composites. The dense PZT-based ceramics exhibit relatively low d_h and g_h coefficients. The low d_h is due to the condition $d_{33} \approx -2d_{31}$ for these ceramics. Values of d_{33} and d_{31} are indicated for PZT-5H as an example; this leads to a low anisotropy factor $-d_{33}/d_{31} \sim 2.16$, which indicates that there is a small volume change under hydrostatic conditions. The high permittivity and low d_h of the PZT-based ceramics lead to the lowest g_h coefficients in Table 5.2. These are some ferroelectric ceramics that exhibit a high anisotropy factor and low permittivity, such as $PbTiO_3$ in Table 5.2 which has a low magnitude d_{31} compared to its d_{33}. This leads to a significantly improved g_h and $d_h.g_h$ coefficient compared to the PZT-based ceramics. All the ceramics are of high density and high stiffness, leading to a high acoustic impedance. The ceramic materials also exhibit a high mechanical quality factor, Q_m, as shown in Table 5.1, which can lead to *ringing* in response to a short pulse signal.

Table 5.2 Data comparing the hydrostatic coefficients and properties of common ferroelectric materials.

Material	ε_{33}^{T}	d_h pC/N	g_h 10^{-3} V m/N	$d_h \cdot g_h$ 10^{-15} m²/N	Density kg/m³
PZT-4 (hard ceramic)	1300	45	4	180	7500
PZT-5H (soft ceramic)	3400	45 $d_{33} = 593$, $d_{31} = -274$	1.5	67	7500
Modified $PbTiO_3$ (ceramic)	203	32 $d_{33} = 47$, $d_{31} = -7.4$	18	570	7500
PVDF (polymer)	10	14	150	2100	1800
Porous PZT (composite)	230	105	55	5575	<7500
$PbTiO_3$–rubber composite (0–3 connectivity)	45	40	100	4000	<7500
PZT rod (15 vol.%)–polymer composite, (1–3 connectivity)	460	268	66	17 700	<7500

Source: [3, 7, 15].

The PVDF ferroelectric polymer exhibits relatively high g_h and $g_h \cdot d_h$, primarily due to its low permittivity; the polymer is also better impedance matched to water due to its low stiffness and low density. The piezoelectric composite systems in Table 5.2 attempt to exploit the high d_{33} coefficients of the ferroelectric ceramics but produce a composite structure that reduces the magnitude of the d_{31} to increase the -d_{33}/d_{31} anisotropy ratio and also reduces the effective permittivity. It can be seen in Table 5.2 that the successful design of a piezoelectric composite structure can lead to the composites producing high d_h coefficients (due to low magnitude d_{31} and high -d_{33}/d_{31}anisotropy), high g_h and very high $d_h{\cdot}g_h$. Since piezoelectric composites often consist of low-density and low-stiffness polymers, they also exhibit a low acoustic impedance and low Q_m compared to the ceramic counterparts.

Note 5.9

The small volume change of ferroelectric ceramics under hydrostatic conditions leads to relatively low d_h coefficients (charge-based hydrostatic sensors). Ferroelectric polymers with low permittivity can provide a high g_h (voltage-based hydrostatic sensors) and $d_h{\cdot}g_h$ (signal-to-noise ratio). The piezoelectric composites provide a route to achieve both a high d_h (due to a low d_{31}) and g_h (due to low permittivity), leading to very high $d_h{\cdot}g_h$.

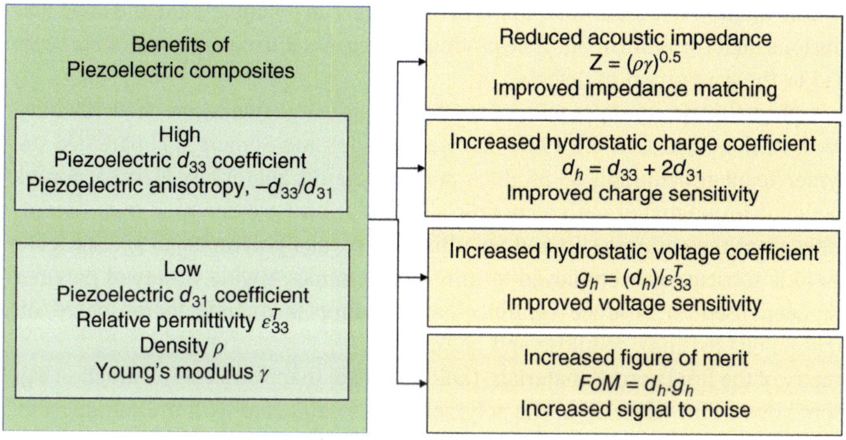

Figure 5.15 Summary of potential benefits of piezoelectric composites.

Figure 5.15 summarises some of the potential advantages of piezoelectric composite structures.

5.6.1 Production of Piezoelectric Composites

There are a number of methods to produce a porous piezoelectric material, and these are summarised in the following section. This will include relatively simple and low-cost approaches that have the potential for large-scale production and more complex routes that are more appropriate for high-value applications.

a) Porous Piezoelectric with (3–3) Connectivity

One approach to producing a porous piezoelectric ceramic is to use the 'BurPS' technique, which stems from the phrase 'burnt-out polymer spheres'; this was first reported by Shrout, Schluze and Biggers [16]. This manufacturing route involves mixing a volatile inclusion material with the piezoelectric ceramic powder, better pressing the powder into the required geometry and eventually high-temperature sintering, which burns out the volatile material to produce the pores. Possible volatile inclusion materials include a polymer, and while the geometry of inclusions is often spherical, the inclusion geometry can be tailored to the exact nature of the pores that are required. During a high-temperature treatment, the volatile inclusion material is removed, where typical temperatures are in the range of 300–500°C. The burn-out of the inclusion then left pores in the material with similar dimensions and shape to the inclusion. After a further heat treatment to produce a sintered ceramic, the microstructure contains pores within the ceramic microstructure. This is a relatively simple approach, and the dimensions and shape of the pores in the piezoelectric ceramic microstructure can be tailored to some extent by selecting inclusions with different dimensions, geometry and volume fractions.

One advantage of the BurPS approach is that it does not need specialist equipment and it is therefore a low-cost, rapid and easy process to perform. Nevertheless, the BurPS approach can produce materials with a limited range of porosity, typically less than 50 vol.% porosity.

For pore volume fraction in excess of 50 vol.%, the material can collapse while the large volume of inclusions melts or is burnt out. The production of gases during the burn-out process can also lead to the generation of defects.

To produce piezoelectric ceramics with over 50 vol.% porosity, alternative methods have been explored. In this regard, in the 1960s, Schwartzwalder and Somers [17] patented the use of polymer foam reticulation to produce porous ceramic materials. Here, a porous polymer sponge is impregnated with a ceramic slip, and a heat treatment is then used to burn off water from the ceramic slip and volatilise the porous polymer sponge. A higher heat treatment is subsequently employed to sinter the ceramic. A wide variety of polymer foams have been used to produce ceramic foams; examples include foams based on poly(urethane), poly(styrene), cellulose and latex.

The geometry of the final porous materials, their pore size, shape and volume fraction can also be tailored by careful selection of the polymer foam to infiltrate with the ceramic slip. There is also potential to achieve near-net shape forming since there is limited shrinkage of a highly porous ceramic on sintering.

Since the process to impregnate the ceramic slip into the polymer foam is the same, irrespective of the pore size and shape of the pores, there is potential to also produce ceramic with a graded level of porosity, for example, to produce functionally graded structures [18].

b) 1–3 Piezoelectric Composites

1–3 piezocomposites consist of piezoelectric ceramic rods (or fibres) that are embedded in an interconnected polymer matrix. To produce a 1–3 composite, the *dice and fill* is often employed and is based on the careful cutting of a sintered dense piezoelectric ceramic. To create the piezoelectric rods (or 'posts'), a series of parallel cuts are produced in the dense piezoelectric ceramic via a precision saw; the ceramic sample is then rotated by 90°, and identical cuts are subsequently made at right angles to the original cuts, as shown in Figure 5.16. As a result of this cutting process, an array of square piezoelectric diced posts is produced; if the cuts do not traverse through the whole ceramic thickness, the diced posts remain attached to a ceramic base to facilitate easy handling, as shown in Figure 5.16.

This diced ceramic article can then be infilled with a polymer to produce the 1–3 piezocomposite; the non-cut base in Figure 5.16 can be removed after impregnation. The structure of the final 1–3 piezocomposite is somewhat limited by the width of the dicing saw employed; this process is also relatively slow and is manually intensive. Fine-scale piezocomposites with 1–3 and 2–2 connectivity patterns can also be produced by a process of ceramic injection moulding. Inkjet printing has also been used to produce three-dimensional ceramic structures, which contain small-scale features [20].

The growth in additive manufacturing provides new approaches to developing piezoelectric composite structures. Zeng et al. [21] used a mask-image-projection-based stereolithography (MIP-SL) process, a 3D printing method. The additive manufacturing approach was employed to build a $BaTiO_3$-based piezoelectric composite that has a honeycomb structure. The advantage of additive manufacturing is the additional design flexibility in terms of potential geometries compared to the dice-and-fill approach described above. After three-dimensional printing in the green state, the ceramic is sintered, which leads to some shrinkage, as shown in Figure 5.17. After forming the ceramic structure, it can then be impregnated with a polymer, such as an epoxy, to create the piezoelectric composite, as shown in Figure 5.18.

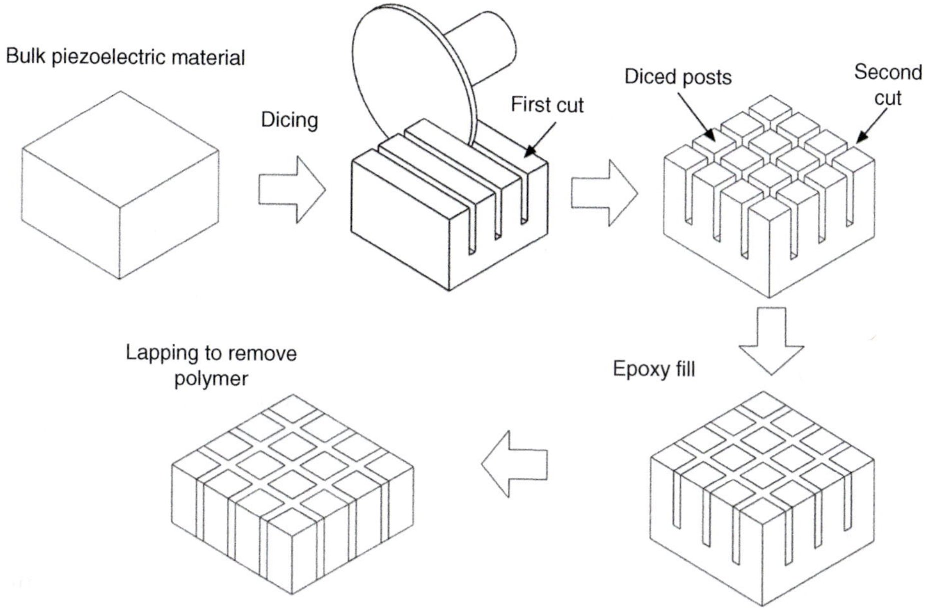

Figure 5.16 Diagram of the dice-and-fill method to create a 1–3 piezocomposite. *Source:* Lee et al. [19] / MDPI / CC BY 4.0.

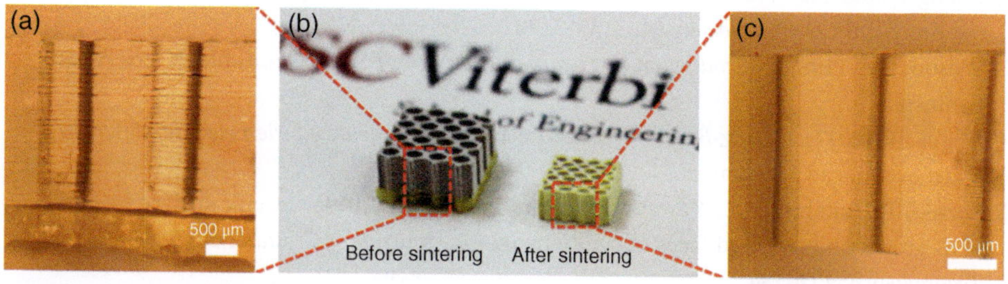

Figure 5.17 Additive manufacturing of piezoelectric materials. (a) Layer details before sintering. (b) Comparison of the size of printed piezoelectric before and after sintering. (c) Layer after sintering. *Source:* Zeng et al. [21] / MDPI / CC BY 4.0.

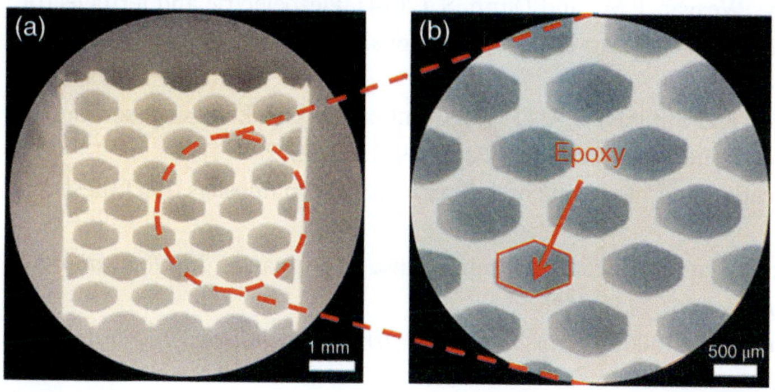

Figure 5.18 (a) Images of a sintered sample that was subsequently filled with epoxy resin; scale bar is 1 mm. (b) Microscope image of the sample with a scale bar of 500 µm. *Source:* Zeng et al. [21] / MDPI / CC BY 4.0.

5.7 Conclusions

This chapter has provided an overview of piezoelectric sensors for sensing acceleration, pressure, force and hydrostatic loads in sonar applications. The range of potential sensor materials in terms of ferroelectric ceramics, ferroelectric polymers, single-crystal piezoelectric materials (quartz) and composite materials has been outlined. Sensing modes in terms of charge sensitivity and voltage sensitivity have been examined, along with environmental factors such as frequency range, thermal drift and operating temperature. Various sensor configurations have been examined, such as d_{33}-mode, d_{31}-mode and shear d_{15}-mode. The impact of hydrostatic conditions in sonar has also been examined. The next chapter aims to accumulate the piezoelectric charge generated during mechanical loading to provide electrical power, rather than a sensing signal; this is the basis of energy harvestsing.

References

1 Serridge, M. and Licht, T.R. (1987). *Piezoelectric Accelerometers and Vibration Preamplifiers: Theory and Application Handbook*. Denmark: K Larden & Son A/S.

2 Zhang, S. and Yu, F. (2011). Piezoelectric materials for high temperature sensors. *Journal of the American Ceramic Society* 94: 3153–3170.

3 Wilson, O.B. (1988). *Introduction to Theory and Design of Sonar Transducers*. Los Altos: Peninsula.

4 Saigua, Y. (2017). *Quartz-Based Piezoelectric Materials, Advanced Piezoelectric Materials*, 2nde. Woodhead Publishing.

5 Suzuki, T., Takahashi, K., Kawasaki, M., and Kagami, T. (2014). Specific surface free energy of as-grown and polished faces of synthetic quartz. *Journal of Crystallization Process and Technology* 4: 177–184.

6 Uchino, K. (2003). *Introduction to Piezoelectric Actuators and Transducers*, International, Center for Actuators and Transducers, 16802. University Park: Penn State University.

7 Berlincourt, D. and Krueger, H.H.A. (2002). *Properties of Morgan electroceramic ceramic*, Technical Publication TP-226,. Morgan Electroceramics.

8 Bowen, C.R., Kim, H.A., Weaver, P.M., and Dunn, S. (2014). Piezoelectric and ferroelectric materials and structures for energy harvesting applications. *Energy and Environmental Science* 7: 25–44.

9 Dues, S.M., Adams, J.M., and Shinkle, G.A. (1990). Combustion knock sensing: sensor selection and application issues. *SAE Technical Paper* 99: 1192–1202.

10 Perry, A. (2002). Fabrication and modelling of 3–3 piezoelectric composites. PhD thesis. University of Bath.

11 Young, J.W. (1977). Optimization of acoustic receiver noise performance. *The Journal of the Acoustical Society of America* 61: 1471–1476.

12 Sherman, C.H. and Butler, J.L. (2007). *Transducers and arrays for underwater sound*. New York: Springer-Verlag.

13 Rathod, V.T. (2020). A review of acoustic impedance matching techniques for piezoelectric sensors and transducers. *Sensors* 20: 4051.

14 Newnham, R.E., Skinner, D.P., and Cross, L.E. (1978). Connectivity and piezoelectric-pyroelectric composites. *Materials Research Bulletin* 13: 525–536.

15 Moulson, A.J. and Herbert, J.M. (2003). *Electroceramics*, 2nde. Chichester: Wiley.

16 Shrout, T.R., Schluze, W.A., and Biggers, J.V. (1979). Simplified fabrication of PZT/polymer composites. *Materials Research Bulletin* 14: 1553–1559.

17 Schwartzwalder, K., Somers, A.V. (1963). Method of making porous ceramic articles. US Patent No. 3 090 04.

18 Cichocki, F. and Trumble, K. (1998). Tailored porosity gradients via colloidal infiltration of compression moulded sponges. *Journal of the American Ceramic Society* 81: 1661–1664.

19 Lee, H.J., Zhang, S., Bar-Cohen, Y., and Sherrit, S. (2014). High temperature, high power piezoelectric composite transducers. *Sensors* 14: 14526–14552.

20 Senlis, G., Dubarry, M., Lejeune, M., and Chartier, T. (2002). 3D piezoelectric structures made by ink-jet printing. *Ferroelectrics* 273: 279–284.

21 Zeng, Y., Jiang, L., Sun, Y. et al. (2020). 3D-printing piezoelectric composite with honeycomb structure for ultrasonic device. *Micromachines* 11 (8): 713.

6

Energy Harvesting

6.1 Introduction

Energy harvesting is attracting increasing interest, as it provides a pathway to create self-powered and autonomous low-power electronic devices; these include wireless sensor networks, wearable electronics, medical devices and the growing demands for large numbers of sensors that are needed for the Internet of Things (IoT) and Industry 4.0. As an example, an energy harvester is commercially available that is able to convert the vibrations of rail rolling stock into electrical power to achieve self-powered wireless communication of sensing data to provide information on the health of rail bearings [1].

This capability to provide a continuous supply of electrical energy to low-power electronic systems using energy harvesting has a range of advantages. Firstly, there is potential to reduce the cost related to the provision of battery-powered supplies, and it significantly reduces the maintenance time, running costs associated with the management of any battery power supplies and the costs related to the installation of cables to provide mains power. The ability to provide a degree of autonomy in terms of delivery of electrical power is of relevance for any sensor networks placed in remote regions or locations that are difficult to reach, or for sensors or communication systems that are being used in hostile environments. This can include safety monitoring devices, micro-sensors which can be attached to structures, medical implants and sensors in industrial environments. Energy harvesting can reduce damage to our environment by reducing the need to dispose of batteries and the materials they contain, which can include heavy metals.

As a result, energy harvesting provides the potential to deliver a *battery-free* solution by being able to scavenge ambient forms of energy; this can include mechanical vibrations, heat, light, water flow and wind. While energy harvesting materials and devices are continuing to improve, there are also advances in microprocessor and wireless communication technologies, leading to reduced power consumption and demand. Energy storage approaches are also continuing to improve; examples include supercapacitors [2] or the relatively recent concepts of structural power or batteries [3]. It is this growing convergence of energy harvesting, energy storage and low-power technologies/wireless communications that can provide significant potential to produce energy harvesting products and systems.

Piezoelectric Materials: Properties, Applications and Devices, First Edition.
Christopher R. Bowen and Andrew R. Plummer.

6.2 Concept of Piezoelectric-Based Energy Harvesting

The harvesting of mechanical energy involves converting mechanical energy from motion, structural deformations, strains, pressure changes or vibrations into usable electrical energy and power. A range of mechanical energy sources is available for harvesting; these include mechanical vibrations from transport and machinery [4], the mechanical strain associated with the contact patch of an automotive tyre as it deforms on contacting the road [5], fluid flow from air or water flow [6, 7], human motion from walking [8], in-body motion, such as the movement of the arm, chest or heart [9], and the forces generated in orthopaedic implants during walking. While the kinetic energy of wind is used as a primary source of power, namely for power *generation* (kW to MW), the concept of energy *harvesting* has a focus on small levels of power by scavenging ambient energy, where batteries, the electrical grid or generators are not an option. While many applications of energy harvesting are for low-power applications, ranging from μW to mW, there is also developing interest in higher power piezoelectric systems operating in the W-to-kW range [10].

In the context of piezoelectric energy harvesting, the two primary approaches to scavenging energy are often subdivided into (i) *inertial harvesting* and (ii) *kinematic harvesting* [11]. The concept of *inertial energy harvesting* involves some form of resistance of a mass to acceleration [11], which leads to the generation of a mechanical load within a mass–spring system during the vibration or motion of the energy source. Inertial systems can often be used in mechanical energy harvesting when they are directly connected to a moving mechanical source of mechanical energy. As the mechanical base source moves, there is also a mechanical vibration within any mass–spring system that is attached to it. As a result, it is possible to harvest that mechanical energy to produce electrical energy by exploiting the *direct* piezoelectric effect. For inertial systems oscillating at their resonant frequency, it is possible for the vibrational amplitude of the system to be much greater than the base movement; the *inertial* approach therefore provides the potential to tune the system response to the motion that is being harvested. This approach is of relevance if the mechanical source has a specific frequency or if vibrations are in a narrow frequency range; a good example would be vibrations associated with helicopter rotors that are rotating at a fixed frequency, e.g. 17 Hz.

The second approach to energy harvesting is termed *kinematic energy harvesting* [11], which aims to couple the mechanical harvester with the motion of a mechanical energy source. Examples of such systems include scavenging energy associated with the deformation and strain of the wall of an automotive tyre for sensors associated with tyre pressure monitoring [12] or the flexure and strain associated with human motion for low-power wearable electronics [13]. As the mechanical deformation and strain experienced by the energy harvester are coupled to a tensile, compressive, bending or shear strain of the mechanical source, the harvester may not be simply connected by a single point to the source; it can be many points or even distributed across a structure. As a result, kinematic energy harvesting relies on the deformation of a structure and is not reliant on inertia or a resonant frequency, unlike inertial harvesting [11].

A typical inertial or kinematic energy harvester therefore aims to extract energy from a moving mechanical energy source and convert it into some form of electric power via the direct piezoelectric effect.

Figure 6.1 Schematic of a common energy harvesting process; V is voltage, and I is current. *Source:* Bowen et al. [11] / with permission of Royal Society of Chemistry.

The concept of energy harvesting is outlined diagrammatically in Figure 6.1, and based on this overall approach, it is of interest to understand the following details of any system:

I) details on the nature of the energy source; this includes vibration/motion amplitude and frequency of the vibration/motion and if the source is fixed frequency or variable frequency,
II) the mechanism by which the source mechanical energy is delivered to the energy harvesting device, for example, whether the harvester is based on *inertial* or *kinematic* approaches,
III) the mechanism by which electromechanical conversion is achieved by the energy harvesting transducer; this can include the direct piezoelectric effect or other mechanisms such as electromagnetic induction or triboelectric effects and
IV) the method by which energy produced by the harvester is delivered to the electrical system that requires power, for example, storage on a capacitor or other energy management approaches.

Note 6.1

Two main approaches for piezoelectric-based harvesting include (i) inertial energy harvesting, leading to a mechanical load on a piezoelectric element within some form of mass–spring system, or (ii) kinematic energy harvesting, where the mechanical energy source and energy harvesting device are more directly coupled.

During the energy harvesting process, a range of potential losses can be incurred; these are not simply restricted to the harvesting device [14] since there are system losses during each stage outlined in Figure 6.1. Examples of system boundary losses include the transfer of electrical energy from the transducer to the electrical load, the method by which the harvester is connected to the mechanical source, the coupling coefficient of the piezoelectric material employed or losses from the storage capacitor. Some of these aspects will be discussed later in the chapter in the derivation of a range of performance figures of merit (FoMs) for piezoelectric materials.

If we consider the concept of vibration energy harvesting, the simplest source of vibration we can consider is a single-frequency sinusoid. This vibration source can be characterised by both a *frequency* (f) and a vibration *amplitude*. The amplitude can be often described via acceleration (a), although we can also use the velocity (v) or displacement

(d) since all of these terms are linked to the vibrational frequency by the equations shown below [11].

$$a = a_0 \, Sin(2\pi f \, t); \; v = -\frac{a_0}{2\pi f} Cos(2\pi f \, t); \; d = -\frac{a_0}{(2\pi f)^2} Sin(2\pi f \, t)$$

where t is time. It is of interest to compare the impact of frequency on the acceleration, velocity and displacement in the three equations above. While many potential vibration sources to be harvested are not simple sinusoids, there are several vibration sources such as rotors, pumps or industrial machinery, which operate with a strong frequency component. Vibration sources are often characterised [15] by their peak amplitude at resonance, although not all sources can be easily described in this way. The growth in energy harvesting means that vibration profiles of a wide range of engineering systems (transport, civil structures, body sources and white goods) are now often collected and made available for the development of energy harvesting systems [16].

There is a wide range of technologies and mechanisms for using motion/vibration in the form of acceleration or strain to harvest and produce electrical power. Electromagnetic generators (EMGs) have been widely used and successfully deployed in commercial systems. However, their efficiency decreases when used in harvesting devices that require relatively small scales/dimensions. As a result, EMGs are therefore considered less appropriate for small-scale systems (mm^3 to cm^3). In contrast, the potential to use piezoelectric materials for energy harvesting is of interest when there is a need to achieve mechanical-to-electrical energy conversion at small volumes, with potential for integration into microelectromechanical systems (MEMS) devices and electronic circuits [11]. As an example, the power density for a piezoelectric transducer can be greater than an EMG for volumes less than ~ 1 cm^3 [17].

As described in Chapters 2 and 3, there will be a change in the polarisation of a piezoelectric material when we apply mechanical stress or strain. Due to the change in polarisation, an electrical charge will be generated to produce an electrical potential and lead to the flow of current in an external electrical circuit; the charge can also be accumulated within a storage capacitor. As a result, piezoelectric materials can be employed as an *electromechanical* generator. Figure 6.2a shows how the initial polarised material is in a stress-free state. Since the material is polarised, screening charges are attracted to the top and bottom electrodes of the material to provide an electrostatic balance. In this state, the system can be considered to be in a state of equilibrium, and no current will flow when connected to an electrical circuit/load. If we now consider the application of a compressive load into the piezoelectric material in the polarisation direction so it is operating in d_{33}-mode, as shown in Figure 6.2b, there will be a decrease in the polarisation of the material. Due to the decrease in polarisation, a fraction of the surface charge will be free to flow within the electrical circuit it is connected to, leading to the generation of a potential difference across the piezoelectric and a current flow in any electrical load. The current level will depend on (i) the amount of charge (Q) that is produced (which depends on the mechanical load and the change in polarisation), (ii) the rate at which the mechanical load is applied (since the current is associated with the rate of current flow, dQ/dt) and (iii) the nature of the electrical load (e.g. a high resistance will act to restrict the current flow). If the compressive load is removed or tensile stress is applied, as shown in Figure 6.2b, the polarisation

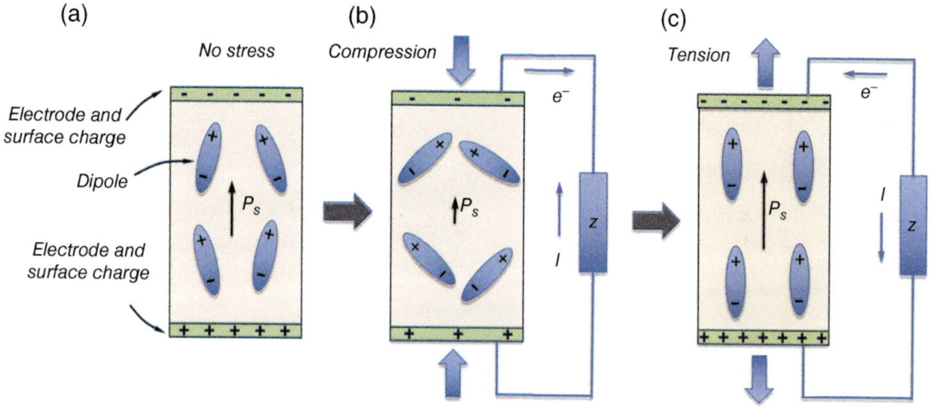

Figure 6.2 Schematic of a piezoelectric material subject to (a) stress-free state and (b) charge generation by a compressive load or (c) tensile mechanical load.

level will increase, which leads to a current flow in the opposing direction to allow screening charges to balance the new state of polarisation and surface charge. Clearly, piezoelectric materials therefore generate an AC output in response to a time-varying stress or strain and will often require some form of rectification if being stored in a capacitor or if DC operation is required.

As shown in Figure 6.2 and described in Chapters 2 and 3, piezoelectric materials typically have a specific polarisation axis/direction, and the amount of charge generated will depend on the direction of the applied load or strain, relative to that polar axis. The notation used is that the direction of polarisation is represented by the 3-direction (or z-direction). For many piezoelectric ceramics, due to the symmetry of the crystal structure (see Chapters 2 and 3), directions in the plane at right angles to the polarisation direction are considered to be equivalent and are termed the '1-direction' (or x-direction); for example, the properties in the 1-direction and 2-direction are the same in many ferroelectric ceramics, such as the lead zirconate titanate (PZT) family of materials. It is therefore of interest to note that it is possible to design a piezoelectric generator, where stress is applied in a direction that is parallel to the polarisation dirsection, as shown in Figure 6.3, or at 90° to the polarisation direction. This results in two different generator configurations that are often employed in piezoelectric generators in Figure 6.3, which are termed 'd_{33}-mode' (stress is parallel to the polarisation axis, Figure 6.3a) and 'd_{31}-mode' (stress at right angles to the polarisation axis, Figure 6.3b); this is the same notation used for piezoelectric actuators and sensors in Chapters 4 and 5. The maximum energy per cycle is also indicated in Figure 6.3, indicating that the energy is proportional to the $d_{ij} \cdot g_{ij}$ product; this will be explored in more detail in this chapter. There is also potential to harvest shear loads in d_{15}-mode to exploit the high d_{15} coefficients of some piezoelectric materials, but these are less well reported [18].

The performance of a material in the context of energy harvesting is highly dependent on the piezoelectric and dielectric properties of the material, particularly the piezoelectric coefficients (e.g. d_{ij} and g_{ij}) and the dielectric coefficients (e.g. permittivity, ε_{33}^{T}). The magnitude of the stress/strain that is applied and the level of coupling between the mechanical energy

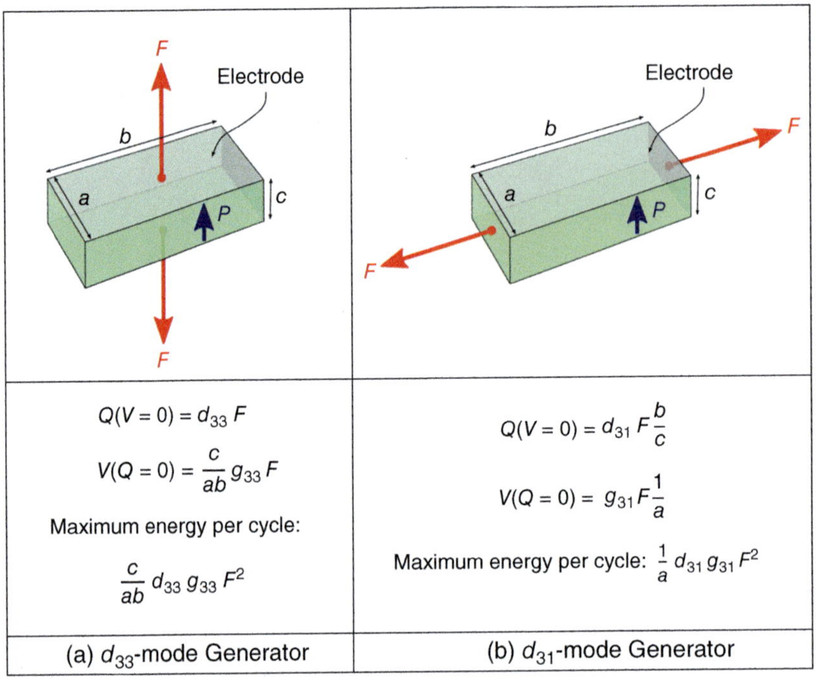

Figure 6.3 Diagram of generators (a) d_{33}-mode and (b) d_{31}-mode piezoelectric generators. Q = charge, V = voltage, F = force, P = polarisation direction, d_{33} and d_{31} = piezoelectric charge coefficients and g_{33} and g_{31} = piezoelectric voltage coefficients. *Source:* Bowen et al. [11] / with permission of Royal Society of Chemistry.

source and the piezoelectric are also important factors in determining the level of energy/power that is produced. The piezoelectric material must also be able to sustain the level of applied stress or strain and must also be able to survive being subjected to a large number of mechanical cycles, with no permanent deformation. For example, if the material is connected to a tyre wall that undergoes significant elastic deformation or strain, the ability to sustain the mechanical stresses and cycles is of particular importance for kinematic energy harvesters since the limit of the mechanical strength and elasticity of a piezoelectric material can influence energy harvesting behaviour; the potential for stress to lead to non-linear deformations due to domain switching has been outlined in Chapter 3 (Section 3.6). This is of importance for many of the inorganic ceramic-based piezoelectric materials, which exhibit low toughness and are unable to sustain large elastic deformations. Again, we will examine in this chapter specific performance FoMs that can consider some of these factors in terms of energy harvesting.

If we now consider an inertial piezoelectric harvester, the strain and deformation of the piezoelectric element are achieved by inertia from a seismic mass that is subject to vibrations and acceleration; this is in contrast to a kinematic harvester, where the stress is produced by being deformed by a mechanical energy source. This can be similar to some of the accelerometer sensor configurations described in Chapter 5 on sensors, although a key difference is that sensors are often designed to operate off-resonance to provide broadband operation, while the inertial energy harvester is often designed to operate at resonance

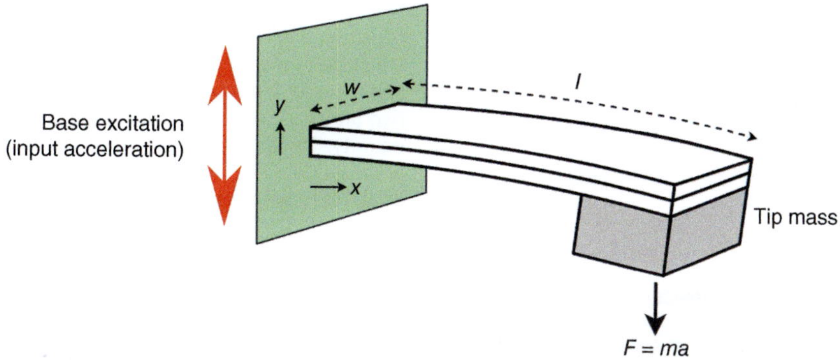

Figure 6.4 Schematic of an inertial piezoelectric-based cantilever vibration energy harvester. *Source:* Bowen et al. [11] / with permission of Royal Society of Chemistry.

to maximise the deflection and power. A common approach to achieving this coupling is to employ a cantilever-based structure, as shown in Figure 6.4.

The cantilever, as indicated by the white volume in Figure 6.4, is mechanically clamped at one end to the source of vibrations, and a tip mass (m) is attached to the unclamped side of the cantilever. When base excitations are applied to the structure (see red arrow in Figure 6.4), there will be an acceleration (a) and an associated velocity (v) and displacement (d), as indicated earlier in this chapter. The inertia associated with the accelerating mass located at the end of the cantilever leads to a deflection of the beam since the resulting force $F = ma$. A simple bending of the piezoelectric material in Figure 6.4 will generate equal, but opposite, strains on the upper and lower surfaces of the beam. If the cantilever was constructed from a single piezoelectric element, it is therefore likely that a combination of tensile and compressive strains would cancel out the generated charges; as a result, there will be no piezoelectric current produced [11]. To use a cantilever structure to produce an energy harvesting generator, there is therefore a need to move the piezoelectric element away from the centre axis of the cantilever beam so that the strain experienced by the piezoelectric element is either fully compressive or fully tensile. This displacement of the piezoelectric element from the neutral axis can be achieved by adhering the piezoelectric element onto a non-piezoelectric and passive elastic layer, thereby leading to the creation of a classical *unimorph* device structure (Figure 6.5a). An alternative approach is to construct a cantilever using two piezoelectric materials, which are polarised in opposing directions (so that the generation of tensile and compressive stresses above and below the neural axis does not lead to a cancellation of charge); this will lead to the formation of a classical piezoelectric *bimorph* structure (Figure 6.5b).

As described, inertial-based energy harvesting generators are typically operated near their resonant frequency, where the amplitude limit of the cantilever beam deflection is related to the overall losses in the mechanical system. The overall loss can include losses due to the fact that some of the mechanical energy is being converted into electrical energy by the direct piezoelectric effect, along with other losses in the system that include frictional loss, dielectric loss and air damping.

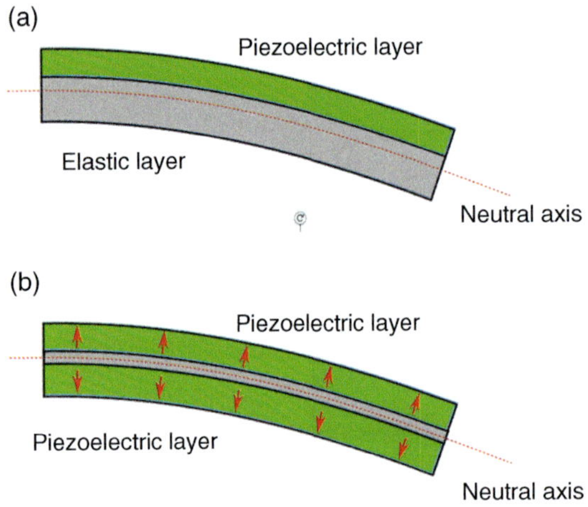

Figure 6.5 Piezoelectric cantilever construction for harvesting: (a) unimorph structure based on a single piezoelectric layer and (b) bimorph structure based on two piezoelectric layers of opposing polarity. The arrows relate to the polarisation direction. *Source:* Bowen et al. [11] / with permission of Royal Society of Chemistry.

We have seen in this book that polycrystalline ceramics, such as PZT, represent an important, and commonly used, family of ferroelectric and piezoelectric materials. As outlined in Chapter 3, these materials contain a number of additives and dopants to tailor their properties for specific applications. This includes compositions based on (i) *hard* ferroelectric materials that exhibit low levels of hysteresis and loss but relatively small piezoelectric charge coefficients due to more limited domain mobility and (ii) *soft* ferroelectric materials that exhibit higher piezoelectric charge coefficients but larger levels of hysteresis and loss due to their more mobile domains. While hard ferroelectric materials exhibit smaller piezoelectric d_{ij} charge coefficients so as to produce a lower level of charge for an applied load, there is potential for them to produce a larger power output compared to soft ferroelectrics due to the lower level of loss and their ability to withstand larger stresses [19]. The hard ferroelectric materials may therefore be attractive in applications where the stress levels are high; this will be examined in the context of hydraulic ripple harvesting as a case study in this chapter. Soft and hard materials will also be compared in terms of potential energy harvesting FoMs within this chapter.

The overall loss within an energy harvesting system can be much larger than the energy that is generated, and efficiency levels in the range of 20% can be common [20]. The mechanical energy source that is being harvested is not an infinite supply, and therefore, converting some of the mechanical energy into electrical energy can lead to damping and a reduction in any vibration; this is another example of a multi-functionality where the material can provide a dual function, namely energy harvesting and passive vibration damping. There are also clear cases where the energy source can be considered to be infinite, for example, using a small energy harvester to harvest the vibrations of a freight train (e.g. for a wireless structural health sensor) to produce mW levels of power will have a negligible impact on the vibration levels of the train as a whole.

To examine the basic mechanics of harvesting and explore potential materials, we will now explore a range of performance FoMs of relevance to the concept of energy harvesting.

6.3 Piezoelectric Properties and Performance Figures of Merit (FoMs)

We have discussed that piezoelectric energy harvesting aims to use the direct piezoelectric effect to produce usable electrical energy from a local mechanical source, for example, to act as a power source for wireless sensors [11]. Various FoMs have been developed to aid in the selection of the optimum material for an energy harvester; a key point is that the FoMs provide a *quantitative* approach to material selection since the performance of the piezoelectric (e.g. energy or power generation) is directly related to the magnitude of the FoM. The FoMs can also be used to design and develop new piezoelectric materials and composites (which are introduced in Chapters 3 and 5) for energy harvesting applications.

A common parameter to assess material for harvesting is the *electromechanical coupling coefficient*, described in Chapter 3. The coupling coefficient quantifies the efficiency of the materials in terms of the conversion of mechanical energy (usually in the form of strain energy) into stored electrical energy [21].

$$k^2 = \frac{\textit{stored electrical energy}}{\textit{input mechanical energy}} = \frac{\textit{stored mechanical energy}}{\textit{input electrical energy}} \tag{6.1}$$

where the electromechanical coupling coefficient is given as

$$k_{ij}^2 = \frac{d_{ij}^2}{\varepsilon_{ii}^X s_{jj}^E} \tag{6.2}$$

where d_{ij} is the piezoelectric charge coefficient, ε_{ii}^X is the permittivity of the material at constant stress, and s_{jj}^E is the mechanical compliance at a constant electric field [11]. It should be highlighted here that the coupling coefficient will only relate to the *efficiency* of the conversion of any mechanical energy into electrical energy. We will see later that this does not necessarily relate to the *magnitude* of the energy that is generated.

As reported in the literature [22, 23], the electromechanical coupling factor does represent a primary factor in determining the harvesting capability of a material. As an alternative, Islam and Priya have derived an 'energy harvesting figure of merit' (FoM_{ij}). If we assume a piezoelectric material is subject to an applied stress that converts strain energy into electrical energy by the direct piezoelectric effect [24–28], the appropriate energy harvesting FoM that should be used to *maximise* the electrical energy produced is

$$FoM_{ij}^X = \frac{d_{ij}^2}{\varepsilon_{ii}^X} \tag{6.3}$$

On comparing Eq. (6.2) (related to conversion efficiency) and Eq. (6.3) (related to maximising energy generated), the difference is that there is no compliance term in Eq. (6.3). This is of interest since polymeric piezoelectric materials often have low coupling coefficients due to their high compliance (leading to low coupling coefficients; see Eq. (6.2)), but they have a high energy harvesting FoM (FoM_{ij}^X) due to a low permittivity. Therefore, there is potential for a polymer-based piezoelectric to be low efficiency, but it ultimately produces more electrical energy than a more efficient material when it attempts to harvest an applied stress; this will be explained in more detail later in this chapter. The performance FoMs related to harvesting will now be derived.

6.3.1 Derivation of Harvesting Figures of Merit (FoMs)

The electrical output delivered by a piezoelectric material will be dependent on a number of factors; this will include (i) the level of mechanical energy often in the form of elastic/strain energy that is applied to the piezoelectric, (ii) how efficient the material is in terms of converting the mechanical strain energy into stored electrical energy (namely the electromechanical coupling coefficient) and (iii) how the stored electrical energy is then converted into electrical power, for example, across an electrical load [11]. Here, we examine each stage of the energy harvesting process and we will outline the derivation of the coupling factor in Eq. (6.2) that relates to efficiency and the energy harvesting FoM in Eq. (6.3) that relates to electrical energy. In addition, we will consider and compare the appropriate FoMs for converting an applied *stress* into electrical energy and an applied *strain* into electrical energy since there are applications where either the stress or the strain is more relevant.

Figure 6.6a shows an example of harvesting at an applied *stress*, where a pressure ripple from a hydraulic system is harvested by coupling a piezoelectric stack to a hydraulic noise-related pressure ripple that is produced in a hydraulic pipeline during operation; this will be considered as a more detailed case study later in this chapter. The example in Figure 6.6b is related to harvesting an applied *strain*, where the kinematic piezoelectric element is attached to the tyre wall, which is subject to a strain when the tyre deforms as it contacts the road; this is often termed 'the contact patch' [5].

6.3.2 Mechanical Energy Input

We will now try to understand the consequences of applying a *stress* or a *strain* to a piezoelectric material in the context of an energy harvesting application. On applying a mechanical stress, the mechanical input is converted into some form of *stored* mechanical energy (often in the form of elastic/strain energy); the degree of conversion will be related to the mechanical impedance of each component [22]. If there is a simple linear relationship between the applied stress (X) and strain (x), and the piezoelectric will obey Hooke's law,

$$U_{mech,in} = \frac{1}{2}xX \tag{6.4}$$

where $u_{mech,in}$ is the mechanical strain energy per unit volume. Here, the stress case is denoted by a superscript X, and the strain case is denoted by a superscript x for the remainder of this Chapter. In simple terms, strain energy can be represented as the energy under the stress–strain plot, as shown in Figure 6.7. Therefore, knowing the Young's modulus (Y) of the material (where $Y = X/x$), for the condition of harvesting an applied stress,

$$U^X_{mech,in} = \frac{X^2}{2Y} \tag{6.5}$$

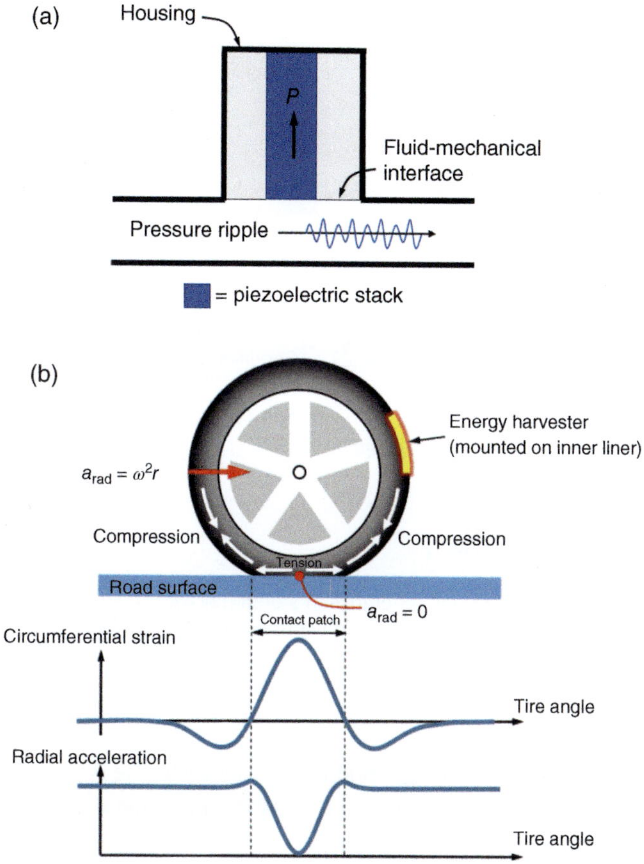

Figure 6.6 Types of piezoelectric energy harvester: (a) stress-driven hydraulic pressure ripple harvester and (b) strain-driven tyre harvester. *Source:* Bowen and Arafa [5] / with permission of John Wiley & Sons.

while for harvesting an applied strain,

$$U^{x}_{mech,in} = \frac{x^2 Y}{2} \tag{6.6}$$

It should be noted that due to the presence of ferroelectric domains, which are mobile under electrical and mechanical stresses, there is potential for these materials to exhibit non-linear stress–strain behaviour, as outlined in Chapter 3. When the harvester is responding to a specific stress, it can be seen from Eq. (6.5) that the stored mechanical energy is inversely proportional to the stiffness (Young's modulus) of the material so that $u_{mech,in} \propto 1/Y$. When the harvester is responding to a specific strain, the condition is different and the stored mechanical energy is proportional to the stiffness, where $u_{mech,in} \propto Y$ in Eq. (6.6). How the material stiffness and applied load influence the mechanical strain energy that is stored in the piezoelectric element is outlined in *Condition* 1 in Figure 6.7; we can observe that when harvesting an applied stress, the stored mechanical energy (namely the strain

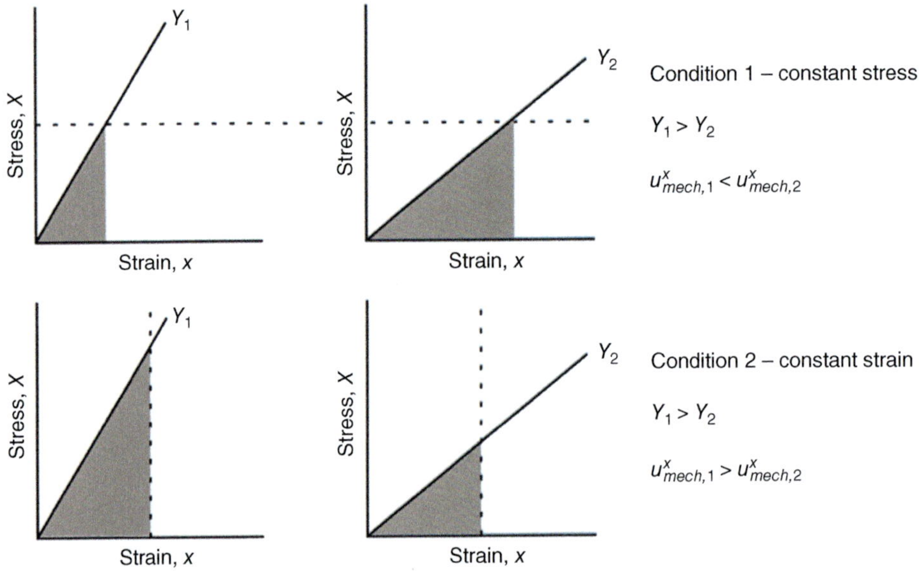

Figure 6.7 Example of linear stress–strain curves and difference in stored mechanical strain energy (area indicated in grey) for a piezoelectric material for high stiffness and low stiffness when harvesting a specific applied stress (indicated by 'Condition 1') or a specific applied strain (indicated by 'Condition 2'). *Source:* Roscow et al. [21] / Springer Nature / CC BY 4.0.

energy) is higher when the material has a low Young's modulus (Y). Piezoelectric polymers are a good example to consider here, which generally exhibit a low Young's modulus compared to piezoelectric ceramics. This low stiffness results in an increase in the stored mechanical strain energy when attempting to energy harvest a specific stress, such as a piezoelectric insole of a fixed area and downward load when it makes contact with the floor. However, if we consider energy harvesting an applied strain, such as the example of attempting to harvest the bending of an automotive tyre wall, as shown in *Condition* 2 in Figure 6.7, a material with a higher Young's modulus will store more mechanical strain energy – in this case, the application of a specific strain to a piezoelectric polymer will lead to a relatively low stress due to its low stiffness. As we progress through this chapter, we also define the elastic properties based on *compliance*, s, which is the inverse of Young's modulus (stiffness), Y, where $s = 1/Y$ [21].

We should examine a range of scenarios where the energy harvester needs to harvest a constant stress (Condition 1) or a constant strain (Condition 2); these have some relation to the inertial and kinematic harvesting approaches outlined earlier in this chapter. More specifically, Crossley et al. specifically described kinematic harvesters as either stress- or strain-driven system [25]. When considering inertial energy harvesting, the piezoelectric material is often joined with a mass that is subject to displacement or vibrations, and the resulting acceleration due to the motion of the mass produces stress in the piezoelectric (Figure 6.4); this is a clear example of Condition 1 in Figure 6.7. Another example of an applied stress (or pressure) and Condition 1 is the hydraulic ripple harvesting in Figure 6.6a. For a kinematic energy harvester, the piezoelectric can be coupled to the displacement of the mechanical energy source; this is an example of Condition 2 (a specific applied strain) in Figure 6.7 such as harvesting strain of a car tyre, as shown in Figure 6.6b. For both stress- and strain-driven approaches, there is also a need to consider the need to achieve mechanical impedance matching between the mechanical source and the piezoelectric material since a highly contrasting stiffness can lead to poor transmission of strain between the source and active materials.

To summarise, there is a need to consider whether the energy harvester that is being developed is either stress-driven (Condition 1) or strain-driven (Condition 2), and to maximise the performance, we should understand that

i) for a stress-driven harvester (Condition 1), it is beneficial to select a piezoelectric material with low Young's modulus and high compliance to maximise the strain energy,
ii) for a strain-driven harvester, it is beneficial to select a piezoelectric material with high Young's modulus and low compliance to maximise the strain energy,
iii) in addition to the points above, good mechanical impedance matching between the mechanical source and piezoelectric material is beneficial.

6.3.3 Converting the Mechanical (Input) into Electrical Energy (Stored)

The next stage considers converting mechanical energy, in the form of strain energy, into electrical energy [21]. For a piezoelectric harvester, the *direct* piezoelectric effect converts the strain energy, $u_{mech,in}$, into stored electrical energy, $u_{elec,stored}$. Piezoelectric materials are not 100% efficient, so there will always be some unconverted strain energy, which we will define as $u_{mech,stored}$. It is not surprising that the amount of unconverted strain energy will be dependent on the electromechanical coupling coefficient of the piezoelectric material, k^2 [22], as described by Eq. (6.1) and in Chapter 3.

$$u_{elec,stored} = k^2 u_{mech,in} \tag{6.7}$$

the 'unconverted' stored mechanical energy is described by

$$u_{mech,stored} = \left(1 - k^2\right) u_{mech,in} \tag{6.8}$$

From these equations above, the stored electrical energy stored, $u_{elec,stored}$, for a given mechanical strain energy input, $u_{mech,in}$, is increased by increasing the k^2 of the piezoelectric. It is of interest to note that if a material exhibits a low coupling coefficient, it may still be the best choice for a stress-driven harvester if the decrease in Young's modulus results in an increase in $u_{mech,in}$ in Eq. (6.5) that outweighs the decrease in k^2. This is due to the fact that, despite the low mechanical-to-electrical energy conversion efficiency of the low-stiffness material, the total amount of strain energy that is available for harvesting is greater in the low-stiffness material.

It is possible to derive piezoelectric energy harvesting FoMs from the stress-driven case (Condition 1) and strain-driven case (Condition 2). If we consider frequency oscillations that are well below typical resonant frequencies (well below 1 kHz), the piezoelectric element can be considered to be a simple parallel plate capacitor [24]. The stored electrical energy can be described by

$$U_{elec,stored} = \frac{1}{2}CV^2 \tag{6.9}$$

where C is the capacitance of the piezoelectric element, and V represents the open-circuit voltage across the capacitor. The capacitance of a piezoelectric element can be determined from

$$C = \frac{A\varepsilon_{33}^X}{t} \tag{6.10}$$

where A is the active cross-sectional area of the piezoelectric element (usually the electrode area), and t represents the thickness of the piezoelectric and the distance between the electrodes [21]. The open-circuit voltage (Vo) that is produced as a result of the application of mechanical stress or strain from the mechanical energy source will depend on the piezoelectric charge coefficient, d_{ij}, and permittivity at constant stress, ε_{33}^X. Let us explore the application of stress in the 3-direction (poling direction), namely X_3.

For a stress (X)-driven harvester,

$$V_{oc} = \frac{d_{33}^2}{\varepsilon_{33}^X} tX_3 \tag{6.11}$$

and for harvesting an applied strain (x),

$$V_{oc} = \frac{d_{33}^2}{\varepsilon_{33}^X} tx_3Y \tag{6.12}$$

If we substitute Eqs. (6.10–6.12) into Eq. (6.9), we can determine the stored electrical energy for both stress-driven and strain-driven cases. It makes sense to divide the total energy by the volume of the piezoelectric element (*area x thickness*) to determine the *specific* energy, namely $u_{elec,stored} = U_{elec,stored}/At$.

For the stress (X)-driven case,

$$u_{elec} = \frac{1}{2}\frac{d_{33}^2}{\varepsilon_{33}^X} X_3^2 \tag{6.13}$$

and for harvesting an applied strain (x),

$$u_{elec} = \frac{1}{2}\frac{d_{33}^2}{\varepsilon_{33}^X} x_3^2 Y^2 \tag{6.14}$$

We have seen that for piezoelectric harvesting materials at stress-driven and strain-driven conditions, the stored electrical energy as a result of stress or strain in the poling is proportional to $d_{33}^2/\varepsilon_{33}^X$; hence, there has been research interest in selecting and designing materials based on this parameter as a stress-driven energy harvesting FoM (see Eq. (6.3)). It can also be observed that Young's modulus of the piezoelectric has no influence on the stored electrical energy for a stress-driven case (see Eq. (6.13)). However, for a strain-driven condition, the stored electrical energy from a strain depends on Young's modulus squared; see Eq. (6.14). This therefore leads to an additional FoM based on maximising the energy density for a strain-driven harvester,

$$FoM_{ij}^{x} = \frac{d_{ij}^{2}Y^{2}}{\varepsilon_{ii}^{X}} \tag{6.15}$$

Instead of using Young's modulus, a strain-driven FoM has been derived that uses the open and closed compliance of a piezoelectric material, namely s_{33}^{E} and s_{33}^{D}, respectively [26],

$$FoM_{ij}^{x} = \frac{d_{ij}^{2}}{\varepsilon_{ii}^{X} s_{jj}^{E} s_{jj}^{D}} \tag{6.16}$$

Note 6.2

It is possible to derive figures of merit to maximise the stored electrical energy for an applied stress or strain, whereby,

i) To maximise $u_{elec,stored}$, there is a need to select materials with a high $d_{ij}^{2}/\varepsilon_{ii}^{X}$ for both stress-driven (Condition 1) and strain-driven (Condition 1) energy harvesting devices.
ii) For a strain-driven energy harvester, there is a need to select materials with low compliance or high Young's modulus to maximise $u_{elec,stored}$ and to maximise the stored electrical energy; see Eq. (6.15).
iii) For a stress-driven condition, the compliance/Young's modulus of the piezoelectric has no influence on $u_{elec,stored}$; see Eq. (6.13).

6.3.4 Producing an Output from the Stored Electrical Energy

Until now, we have examined (i) the influence of the compliance and Young's modulus of the piezoelectric materials on the conversion of the mechanical input into stored mechanical strain energy and (ii) how a fraction of the stored strain energy is converted into stored electrical energy using the direct piezoelectric effect and the influence of the electromechanical coupling coefficient. However, electrical power has yet to be extracted from the energy harvester. This process of extracting the stored electrical energy is the final stage of an energy harvesting system, as outlined in Figure 6.1 [21]. The nature of the electrical output will depend on the electrical impedance of the harvesting circuit, and the electrical impedance of the piezoelectric element and circuit should be matched to achieve an optimum power output. For example, if we use a simple resistor as an electrical load, the electrical impedance of the resistive load (R_{load}) will need to match the electrical impedance of the piezoelectric element ($1/\omega C_{piezo}$); this is achieved at the condition $\omega C_{piezo} R_{load} = 1$, where ω is the angular frequency.

The maximum electrical energy output of an energy harvester for a specific stored mechanical strain energy is related to the *transmission coefficient*, $\lambda_{\max}$ [22, 23], where the full derivation of the transmission coefficient can be obtained from [27]:

$$\lambda_{max} = max\left(\frac{output\ electrical\ energy}{input\ mechaical\ energy}\right) = max\left(\frac{u_{elec,out}}{u_{mech,in}}\right) \tag{6.17}$$

where

$$\lambda_{max} = \left(\frac{1}{k} - \sqrt{\frac{1}{k^2} - 1}\right)^2 \tag{6.18}$$

The 'transmission coefficient' enables the estimation of the maximum electrical energy output as a result of the mechanical energy input by rearranging Eq. (6.17) [21]:

$$u_{elec,out,\,max} = \lambda_{\max} u_{mech,in} \tag{6.19}$$

If we substitute in Eq. (6.7), it can be seen that the electrical energy that is delivered as an output will be proportional to the stored electrical energy, namely the FoMs that we have recently derived, Eqs. (6.3) and (6.16):

$$u_{elec,out,\,max} = \frac{\lambda_{max}}{k^2} u_{elec,stored} \tag{6.20}$$

The maximum electrical energy that is output, $u_{elec,out,max}$, as seen in Eq. (6.20), is proportional to both $\lambda_{\max}/k^2$ and $u_{elec,stored}$ The influence of the coupling coefficient, k^2, on the transmission coefficient, $\lambda_{\max}$ (blue line), and $\lambda_{\max}/k^2$ (dashed green line) is outlined in Figure 6.8, and when $k^2 < 0.5$, which applies to many piezoelectric materials, the value of $\lambda_{\max}/k^2$ does not change significantly. The priority should therefore be to maximise $u_{elec,stored}$ so that the FoMs based on FoM_{ij}^{X} (Eq. (6.16)) and FoM_{ij}^{x} (Eq. (6.17)) are of most interest.

In this section, we have considered the primary stages of energy harvesting; these include (i) applying a mechanical input, often in the form of some strain energy, into a piezoelectric element/material; (ii) converting the mechanical strain energy into stored electrical energy using the direct piezoelectric effect; and (iii) converting the stored electrical energy into some usable electrical energy, namely producing an electrical power output.

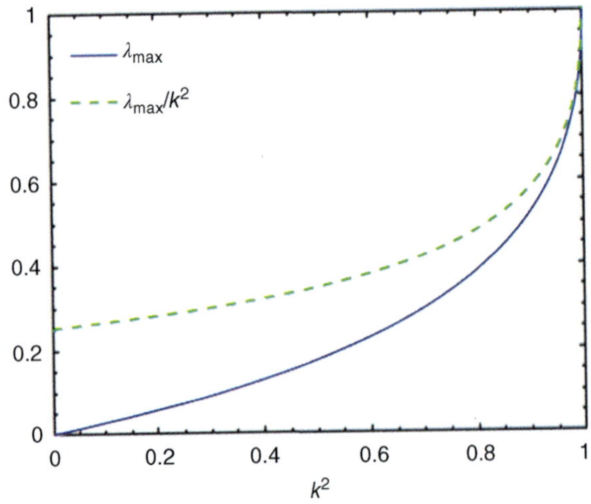

Figure 6.8 Graph showing the effect of the electromechanical coupling coefficient, k^2, on the transmission coefficient, both $\lambda_{\max}$ and $\lambda_{\max}/k^2$. *Source:* Roscow et al. [21] / Springer Nature / CC BY 4.0.

Note 6.3

To summarise the relevant points above for off-resonance harvesting:

i) A high k^2 provides a high transmission coefficient, λ_{max}, and λ_{max}/k^2,
ii) For the condition $k^2 < 0.5$, which is the case in a large number of ferroelectric materials, the aim should be to maximise $u_{elec,stored}$, as there are no large changes in λ_{max}/k^2 for this condition. Therefore, the important performance FoMs for stress-driven or strain-driven harvesting are FoM_{ij}^{X} (Eq. (6.3)) and FoM_{ij}^{x} (Eq. (6.15)).

By now, we have considered the range of performance FoMs (e.g. Eqs. (6.2, 6.3, 6.16 and 6.20)), and the complexities associated with the design and selection of piezoelectric in both bulk and composite forms. Figure 6.9 represents this process graphically, where key material properties, equations and FoMs are outlined. The first stage in Figure 6.9 represents the mechanical source, where some of the mechanical energy is transferred into the

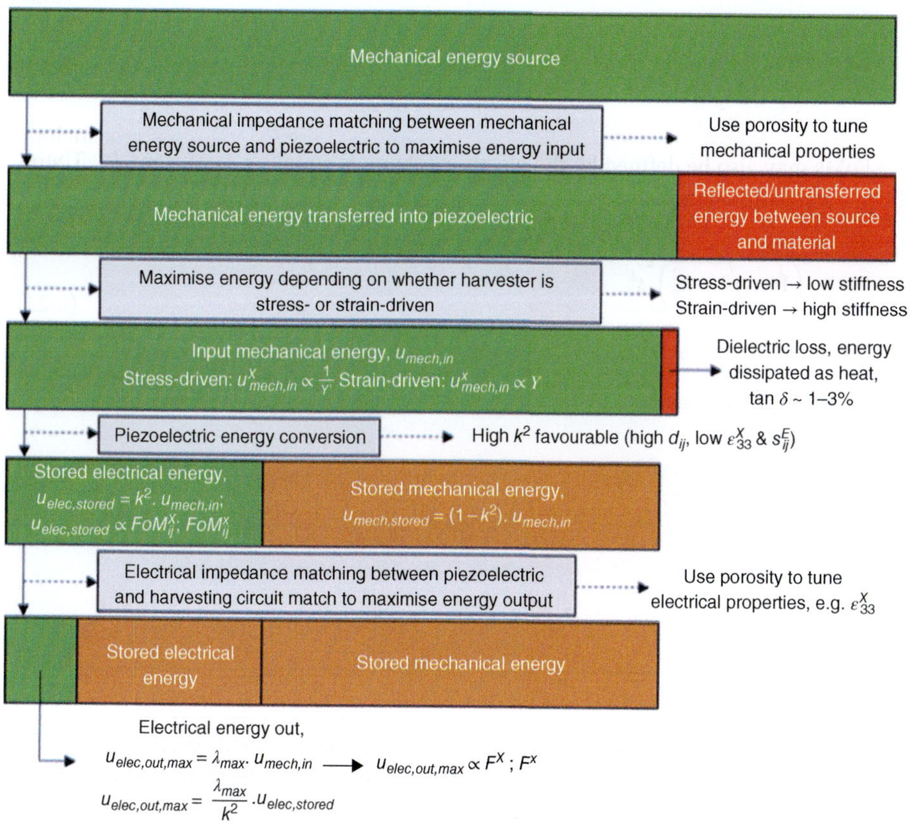

Figure 6.9 Schematic of considerations for design and selection of piezoelectric materials at each stage of the energy harvesting process. *Source:* Roscow et al. [21] / Springer Nature / CC BY 4.0.

piezoelectric as strain energy; any remaining energy is usually reflected depending on how well the piezoelectric device and the source are mechanically impedance matched. Some mechanical energy can be lost as heat due to domain motion or dielectric loss ($\tan \delta$), as outlined in Chapter 3. The mechanical strain energy can then be converted into stored electrical energy [21–23], which is related to the energy harvesting FoMs (Eqs. (6.3 and 6.16)). Finally, the stored electrical energy is used to provide power to an electrical load, depending on the transmission coefficient.

Let us now examine the whole energy harvesting process, as outlined diagrammatically in Figure 6.9. We can define two final piezoelectric harvesting FoMs by substituting the two FoMs that we have derived from the stored electrical energy for (i) a stress-driven process and (ii) a strain-driven process into Eq. (6.20).

First, for harvesting an applied stress:

$$F_{ij}^{X} = \frac{\lambda_{max}}{k_{ij}^{2}} \frac{d_{ij}^{2}}{\varepsilon_{ii}^{X}} \tag{6.21}$$

and for harvesting an applied strain:

$$F_{ij}^{x} = \frac{\lambda_{max}}{k_{ij}^{2}} \frac{d_{ij}^{2}}{\varepsilon_{ii}^{X} s_{jj}^{E} s_{jj}^{D}} \tag{6.22}$$

The FoMs may also be defined by the electromechanical coupling coefficient, k^2. Therefore, for a stress-driven condition,

$$F_{ij}^{X} = \frac{\left(1-\sqrt{1-k_{ij}^{2}}\right)^{2} s_{jj}^{E}}{k_{ij}^{2}} \tag{6.23}$$

and for a strain-driven case,

$$F_{ij}^{x} = \frac{\left(1-\sqrt{1-k_{ij}^{2}}\right)^{2}}{k_{ij}^{2} s_{33}^{D}} \tag{6.24}$$

Note 6.4

The stiffness of the piezoelectric element has a strong influence on the electrical energy that is generated. For instance, for a condition of constant stress, the strain energy in the piezoelectric material can be increased by selecting a material with a low Young's modulus. When a specific strain is applied, the strain energy can be increased by selecting a material with a high Young's modulus.

Note 6.5

The selection of piezoelectric materials with a high electromechanical coupling coefficient provides a more efficient conversion of the strain energy into stored electrical energy. For a stress-driven condition, a material with a low electromechanical coupling coefficient, namely low efficiency, materials with a low stiffness can still produce a larger electrical output than a high-coupling-coefficient material of high stiffness due to the larger strain energy in the low-stiffness material.

Note 6.6

From Eqs. (6.3, 6.16, 6.21 and 6.22), a high $d_{ij}^2/\varepsilon_{ii}^X$ is beneficial to the electrical energy output, irrespective of operating under stress-driven or strain-driven condition.

This section has shown that an understanding of a whole system is required for the effective selection of the piezoelectric material. Knowledge of whether an energy harvester is stress-driven or strain-driven will change whether a high- or low-stiffness material should be selected.

To date, we have not considered the maximum stress or strain that the material can withstand during operation. We therefore now consider a case study of a stress-driven pressure ripple energy harvester, as originally outlined in Figure 6.6b.

6.4 Case Study: Piezoelectric Hydraulic Ripple Energy Harvesting

We will now consider a specific case of a hydraulic-based energy harvester. In hydraulic systems, monitoring of industrial processes is needed for parameters, which include vibrations, operating pressure and temperature as a measure of system health. Hydraulic-based harvesting is of interest as a result of the high levels of mechanical energy associated with pressure fluctuations and the flow of hydraulic liquid [28]. In particular, in fluid systems operating at relatively large levels of pressure, noise and pressure ripples are often produced. Energy harvesting of these high-energy ripples is therefore attracting interest to provide the mW levels of power to enable the operation of wireless sensors that aim to achieve health monitoring.

Examples include a system [29] based on a piezoelectric multilayer stack, as shown in Figure 6.10, where a metal diaphragm provides a means to transfer the mechanical energy from the fluid to apply a mechanical load to the piezoelectric stack. The dimensions of the piezoelectric element, particularly area (A), could be used to tailor the maximum level of stress generated by the pressure ripples; see Figure 6.10. Since a nine-piston axial pump was used that was operated at 1500 rpm, the fundamental frequency (f) can be determined

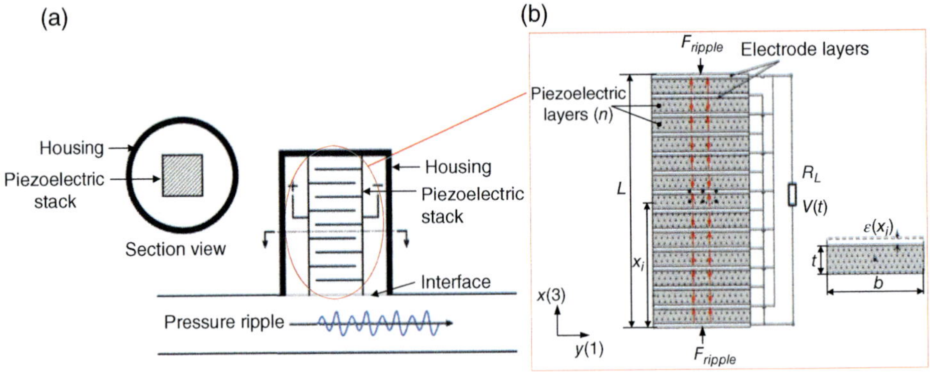

Figure 6.10 (a) Hydraulic pressure ripple harvester and (b) piezoelectric stack and its multilayer architecture; there are n piezoelectric layers; V = voltage, R_L = load resistance, F_{ripple} = oscillating ripple force, L = stack length, and t = thickness of each piezoelectric layer. *Source:* Xiao et al. [28] / with permission of John Wiley & Sons.

as $f = (1500 \times 9)/60 = 225$ Hz [28]. The device was able to deliver a power of 1.2 mW with an oscillating ripple pressure of ~0.4 MPa. Since the piezoelectric stacks exhibit piezoelectric resonances (as outlined in Chapter 3), in the kHz range, this form of harvester is *non-resonant*. In addition, since the piezoelectric harvesting element is directly coupled to the mechanical source, namely the pressure ripple, via a metal diaphragm, it can be considered as a *stress-driven* and *kinematic* energy harvesting system.

Work on ripple harvesting has examined various piezoelectric materials and devices, for example piezoelectric polymers [30], piezoelectric ceramics [29, 31] and ferroelectrics in single-crystal form [32, 33]. In most examples, a multilayer PZT stack has been selected as the piezoelectric device architecture, as shown in Figure 6.10b, which has been formed by combining multiple layers of the piezoelectric ceramic; this is achieved by bonding many ceramic elements or by screen printing multiple layers of piezoelectric ceramic. The multilayer piezoelectric here is considered to be formed from a number of piezoelectric elements, which are connected electrically in parallel. We will examine later how the presence of a number of electrodes influences the charge, current, voltage and electrical energy when the load from noise/ripples is applied.

Previous work [34] has examined the response of the multilayer stack element employed for energy harvesting. Based on our discussions earlier in this chapter, since the ripple load is applied in a direction that is parallel to the polarisation direction (the 3-direction), the piezoelectric element can be considered to be operating in d_{33}-mode; see Figure 6.3. There, the d_{33} piezoelectric charge coefficient is used to estimate the level of charge produced from the ripple load (F_{ripple}).

We now examine a piezoelectric element that has an area, A_s, and length, L, and is manufactured from n layers of piezoelectric material. The thickness of a single piezoelectric layer is t. The material is poled through the layer thickness, namely along the thickness of the stack. We will assume that the operating frequency of the hydraulic system is much lower than the piezoelectric resonance frequency of the stack so that we are operating off-resonance. Note, the case $n = 1$ is analogous to using a bulk material (rather than a stack), and we can use this to examine the influence of using a multilayer structure on the charge, current, voltage and electrical energy [28].

As a result of hydraulic noise, a ripple load F_{ripple} is applied to the piezoelectric element, as indicated in Figure 6.10. The resulting piezoelectric charge (Q) that is produced by the direct piezoelectric element can be calculated from

$$Q = nd_{33}F_{ripple} \tag{6.25}$$

where d_{33} represents the piezoelectric charge coefficient (C/N) of the material. From Eq. (6.25), to produce a large level of charge per unit ripple force (Q/F_{ripple}), we should consider materials with a large d_{33} coefficient and/or select a stack with a high number of layers (n).

As piezoelectric materials are often good insulators of electricity, we can consider the materials as a 'dielectric' and the stack as a 'capacitor'. If we first consider one layer in the piezoelectric stack, its capacitance (C) can be calculated via

$$C = \frac{A_s\varepsilon_{33}^T\varepsilon_0}{t} \tag{6.26}$$

where ε_{33}^T is the relative permittivity at a condition of constant stress, and ε_0 is the permittivity of free space. Since the stack is formed from n individual layers, the capacitance of the stack (C_{eq}) is given by

$$C_{eq} = \frac{nA_s\varepsilon_{33}^T\varepsilon_0}{t} \tag{6.27}$$

As $t = L/n$, the capacitance of the stack can also be given as

$$C_{eq} = \frac{n^2A_s\varepsilon_{33}^T\varepsilon_0}{L} \tag{6.28}$$

The open-circuit voltage (V_{oc}) that is produced due to the charge generated by the ripple load is determined using $Q = C_{eq}V_{oc}$. From Eqs. (6.25) and (6.28), the open-circuit voltage due to the piezoelectric charge is given by [28],

$$V_{oc} = \frac{d_{33}}{\varepsilon_{33}^T\varepsilon_0}\frac{F_{ripple}}{A_s}\frac{L}{n} \tag{6.29}$$

From Eq. (6.29) for a system with a specific ripple load and multilayer stack structure (namely a specific A_s, L and n), the voltage can be maximised by producing a stack based on materials with a high $g_{33} = d_{33}/\varepsilon_{33}^T\varepsilon_0$; this is the *piezoelectric voltage constant* (g_{33}) and is often used to select materials for sensing applications since there is a need to select materials that generate a large voltage per unit stress (see Chapter 5).

In terms of optimising the piezoelectric stack, Eq. (6.29) indicates the voltage can be maximised by reducing the stack area, a low number of piezoelectric layers or a large overall length. Selecting a low number of layers or a large length acts to increase the thickness of the individual piezoelectric layers, t, and a low C_{eq}; see Eq. (6.28), which produces a large voltage, since $Q = C_{eq}V_{oc}$.

The electrical energy, E, that is produced by the oscillating ripple is a better indicator of potential power levels that can be produced. The stored electrical energy (E) in the capacitor is described as [28],

$$E = \frac{1}{2} C_{eq} V_{oc}^{2} = \frac{Q^2}{2C_{eq}} \tag{6.30}$$

which results in

$$E = \frac{1}{2} \frac{F_{ripple}^2}{A_s} \frac{d_{33}^2}{\varepsilon_{33}^T \varepsilon_0} L \tag{6.31}$$

The electrical energy, E, is the electrical energy produced for a single ripple cycle. Since the power $P = Ef$, increasing the electrical energy per cycle, E, or operating/ripple frequency (f) can increase the power that can be produced. Using Eq. (6.31), if we have a specific mechanical source (F_{ripple}) and are using a specific multilayer stack geometry (A_s and L), to increase the energy, we should select materials with a high harvesting figure of merit, $FoM_{33}^X = d_{33}/\varepsilon_{33}^X$; see Eq. (6.3).

Priya [35] has also taken into consideration the impact of dielectric losses by incorporating the dielectric loss of the piezoelectric, which results in

$$FoM_{ij}^X = \frac{d_{ij}^2}{\varepsilon_{ii}^X (\tan \delta)} \tag{6.32}$$

If the material to be used in the piezoelectric element is known, we can also try to optimise the stack geometry. Equation (6.31) indicates that a stack with a low area and long length can increase the energy per cycle; this increase in energy stems from a reduction in the capacitance of the stack. Interestingly, the number of layers (n) to manufacture the stack has no impact on the electrical energy produced per cycle; the number of layers has mainly an effect on the piezoelectric charge (Eq. (6.25)) and voltage (Eq. (6.29)). This stems from the conditions that the capacitance $C_{eq} \propto n^2$, the charge $Q^2 \propto n^2$, and open-circuit voltage $V_{oc}^2 \propto 1/n^2$, where the number of layers in the stack cancels out with regard to electrical energy, since $E \propto CV^2$ [28].

While the use of multiple layers (n) does not increase the electrical energy per load cycle, they have been often employed for ripple harvesting. The potential practical advantages of using a stack include (i) it is possible to combine many layers to increase the stack length and power since the electrical energy is proportional to length (Eq. (6.31)); clearly, producing long bulk ceramics, polymers or single crystals is more difficult and is also difficult to pole, (ii) a large number of piezoelectric layers reduce the voltages that are generated while ripple harvesting, as$V_{oc} \propto 1/n$; this reduces possible electrical damage to any power management system, and (iii) a large number of layers increase the total capacitance, resulting in a small electrical impedance ($Z = 1/j\omega C_{eq}$), which has potential to make impedance matching with an electrical circuit easier.

Since hydraulic systems typically operate at high levels of pressure, there is a need to consider the impact of a large operating stress on the piezoelectric multilayer stack; clearly, the strength of the material provides an indication of the minimum stack area for a known ripple load ($\sigma = F_{ripple}/A_s$). The stress on a piezoelectric material [34] should be below the maximum load that can be applied while preventing any damage; the maximum stress (the 'strength' of the piezoelectric) can be related to the fracture strength, a known fatigue limit (σ_f) or a 'coercive stress' ($\sigma_{coercive}$), where domain switching of the material as a result of the stress leads to permanent depolarisation of the material; this has been discussed in Chapter 3. If we wish to harvest a known ripple force, $F_{\max}$, and mechanical failure/fatigue/depolarisation of the piezoelectric material is to be avoided, we can define a maximum load based on a nominal 'strength' ($\sigma_{\max}$),

$$F_{\max} \leq \sigma_{\max} A_s \tag{6.33}$$

For a known ripple force, $F_{\max}$, the area of the piezoelectric element will be determined by $\sigma_{\max}$. By substituting Eq. (6.34) into Eq. (6.31), it is possible to eliminate the variable related to the stack area, A_s, so that

$$E_{\max} = \frac{1}{2} \frac{\sigma_{\max} d_{33}^2}{\varepsilon_{33}^T \varepsilon_0} F_{\max} L \tag{6.34}$$

As a result, from Eq. (6.35), for a known ripple force $F_{\max}$ and a piezoelectric stack with a fixed length, L, but a variable area, A_s, the material properties of interest to maximise the electrical energy are related to the *ripple FoM*, namely

$$FoM_{ripple} = \frac{\sigma_{\max} d_{33}^2}{\varepsilon_{33}^T \varepsilon_0} \tag{6.35}$$

This particular FoM is able to consider both the energy harvesting performance of the piezoelectric material, in terms of optimising the electrical energy for a ripple load, and the ability of the material to survive mechanical stress.

It is of interest to note that *soft* PZT materials are often of interest for energy harvesting, which is primarily due to their large piezoelectric d_{33} charge coefficients due to mobile ferroelectric domains (see Chapter 3). This can be shown in Table 6.1, where PZT-5H (a well-known soft PZT) exhibits large piezoelectric d_{33} charge coefficients and large coupling coefficients. Since the ferroelectric domains are relatively mobile in these materials, they can be depolarised at relatively low levels of stress; see Chapter 3 and Section 3.6. In contrast, although the more limited domain motion in the hard PZT materials leads to lower piezoelectric d_{ij} charge coefficients (e.g. PZT-4 in Table 6.1), these harder materials exhibit a larger failure strength and higher coercive stress. In addition, the lower domain mobility also leads to a lower permittivity. As a result, from Eq. (6.36), the ability of the hard PZTs to withstand larger stress levels and their lower permittivity makes them interesting for harvesting at high-stress levels, such as hydraulic ripple harvesting.

For a hydraulic application, the maximum operating temperature is also of interest, where the Curie point (T_c) indicates the maximum operating temperature since heating a piezoelectric above its T_c results in the piezoelectric properties being lost [43], as outlined in Chapter 3. A recommended maximum temperature is half the T_c (in °C), for example, a soft PZT with a $T_c = 180\,°\text{C}$ can operate at 90 °C.

Table 6.1 provides a summary of the materials' properties and FoMs for a number of piezoelectric materials; based on the FoMs, we have derived the properties of the key materials including the piezoelectric charge coefficient (d_{33}), dielectric permittivity and loss (ε_{33}^T and

Table 6.1 Summary of properties, including lead zirconate titanate (PZT), lead magnesium niobate–lead titanate (PMN–PT) and lead indium niobate–lead magnesium niobate–lead titanate (PIN–PMN–PT).

Materials	PZT-5H (soft type)	PZT-5A (soft type)	PZT-4 (hard type)	TRSX2B PMN–PT	PIN–PMN–PT (MPB)
d_{33} (pC/N)	593 [36]	374 [36]	289 [36]	1860 [37]	2740 [38]
Relative permittivity at constant stress, (ε_{33}^T)	3400 [39]	1700 [39]	1300 [39]	5000 [40]	7240 [38]
g_{33} (10^{-3}Vm/N)	19.7	24.8	26.1	42.0	42.8
Curie temperature (T_c) or T_{rt} (°C)	$T_c = 193$ [39]	$T_c = 365$ [39]	$T_c = 328$ [39]	$T_{rt} = 85$–98 $T_c > 142$ [40]	$T_c = 197$ $T_{rt} = 96$ [38]
Dielectric loss (tan δ)	0.02 [39]	0.02 [39]	0.004 [39]	≤0.006 [40]	—
Coercive stress (MPa)	20 [41]	20 [39]	82 [39]	—	—
Coercive field (kV/cm)	5.5 [42]	11.8 [42]	14.4 [42]	1.9–2.5 [40]	5.5 [38]
FoM_{33} (Eq. (6.12)) (10^{-12} m^2/N)	11.69	9.35	7.26	78.18	117.17
FoM_{33}/tan δ (Eq. (6.13)) (10^{-14} m^2/N)	5.84	4.65	18.15	130.31	—
FoM_{ripple} (Eq.(6.18)) (10^{10})	2.34	1.86	5.95	—	—
U_{max} (J/m^3)	2337	1859	24406	—	—
k_{33}^2 coupling coefficient	0.56 [39]	0.50 [39]	0.49 [39]	0.83–0.90 [40]	0.90 [38]

Source: Xiao et al. [28] / with permission of John Wiley & Sons.

tan δ), the piezoelectric voltage coefficient (g_{33}) and Curie point (T_c). The FoMs of FoM_{33}^X (Eqs. (6.3 and 6.32)) and FoM_{ripple}, based on the material strength limit (Eq. (6.36)), can be seen.

The range of piezoelectric materials in Table 6.1 has been examined for ripple harvesting, for example (i) polycrystalline soft PZT ceramics (PZT-5H [29, 31, 44, 45] and PZT-5A [46]) and (ii) single-crystal-based materials (lead magnesium niobate–lead titanate [PMN–PT] [32, 33]). The soft PZT materials have large FoM_{33} harvesting FoMs as a result of the relatively large piezoelectric d_{33} charge coefficients. Consideration of the dielectric loss (tan δ) does not make a significant difference in terms of identifying good materials (Eq. [6.32]), as most piezoelectrics have a small electrical conductivity. The hard ferroelectric material (such as PZT-4) exhibits a higher coercive stress due to the limited mobility of domains in these materials. Due to their ability to withstand large stress levels, they have a large ripple figure of merit (FoM_{ripple}) in Table 6.1, over twice as high as the soft PZT-5H ceramic.

The single-crystal-based ferroelectrics in Table 6.1 may be able to provide very large harvesting FoMs. They do have some disadvantages; however, (i) single crystals are often expensive to produce, (ii) it is difficult to produce crystals in complex shapes, (iii) they have low Curie points (or other phase transformations), and (iv) domains are highly mobile and so they have a low coercive stress, although there are currently limited data.

6.5 Pyroelectric Materials and Thermal Energy Harvesting

The previous sections have focused on harvesting mechanical stress or strain because of the change in the polarisation of a piezoelectric with an applied mechanical load. A change in temperature can also lead to a change in polarisation, which is the *pyroelectric effect*. In terms of thermal energy harvesting, heat/thermal energy remains an abundant ambient source of energy that is often wasted. Pyroelectric harvesting has been considered by several researchers [47–49], although thermoelectric materials have been widely considered to harvest temperature gradients (dT/dx, here T = temperature and x = position) via the Seebeck effect. The topic of *pyroelectric thermal harvesting* aims to convert a change in temperature (dT/dt, here t = time) into electrical energy. There are analogies here with piezoelectric energy harvesting outlined previously that convert vibrations ($d\sigma/dt$, here σ = stress) into electrical energy.

We now consider a pyroelectric element that is irradiated by a light source with a power density of W (J/s/m^2) for some time, Δt, that produced a temperature rise in the pyroelectric materials, ΔT (Figure 6.11), which is related to the power density of the light source by [50],

$$\Delta T = \frac{W \cdot \Delta t}{c_E \cdot h \cdot \Delta T} = \frac{W \cdot \Delta t}{c_p \cdot \rho \cdot h \cdot \Delta T} \tag{6.36}$$

where h = thickness (m), c_p = specific heat capacity (J/kg/K), ρ = density (kg/m^3), and c_E = volume specific heat of the material (J/m^3/K). During a temperature rise, where $dT/dt > 0$, there is a decrease in the level of polarisation [51]. The reduction in polarisation results in free charges on the surface of the pyroelectric element. The pyroelectric

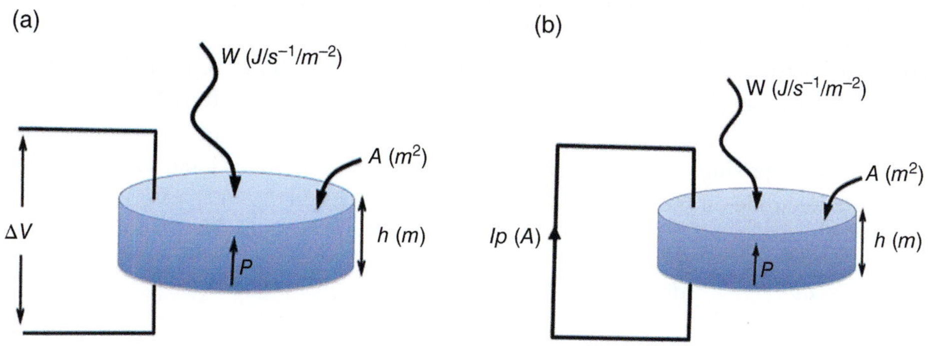

Figure 6.11 Heated pyroelectric at (a) open circuit and (b) closed circuit. The polarisation (P) direction indicated, normal to the top and bottom electrode faces. *Source:* Bowen et al. [50] / with permission of Elsevier.

coefficient, when the material is not clamped, for a constant stress and electric field, is a measure of the charge produced per unit area per unit temperature increase (C/m^2/K) and can be described as

$$p^{\sigma,E} = \left(\frac{dP_s}{dT}\right)_{\sigma,E} \tag{6.37}$$

where P_s is the polarisation, and σ and E correspond to a condition of constant stress and field, respectively [51].

If a pyroelectric material is irradiated via a thermal source with a power density, W, the temperature rise (ΔT) will produce pyroelectric. The amount of charge (ΔQ) generated can be determined by

$$\Delta Q = p \cdot A \cdot \Delta T \tag{6.38}$$

As a result,

$$\Delta Q = \left[\frac{p}{c_E}\right] \cdot \left[\frac{A}{h}\right] \cdot W \cdot \Delta t \tag{6.39}$$

When at an open circuit, the pyroelectric charge produced due to a reduction in polarisation remains to produce a voltage (ΔV) across the electrodes. Since $\Delta Q = C \cdot \Delta V$, where C is capacitance (C), and based on Eqs. (6.26) and (6.39), the open-circuit voltage across the pyroelectric element as a result of the temperature change is [50]:

$$\Delta V = \frac{p}{\varepsilon_{33}^{\sigma}} \cdot h \cdot \Delta T \tag{6.40}$$

Also,

$$\Delta V = \left[\frac{p}{c_E \cdot \varepsilon_{33}^{\sigma}}\right] \cdot W \cdot \Delta t \tag{6.41}$$

where $\varepsilon_{33}^{\sigma}$ is the permittivity at constant stress (F/m).

If we use a short circuit, as outlined in Figure 6.11b, the temperature increase results in the generation of a current that flows in the circuit connected to the upper and lower electrodes. Equation (6.42) provides the closed circuit (i_p), which we can see is dependent on (i) the pyroelectric coefficient (p), since it relates to the level of charge (Q), and (ii) the rate of change of temperature (dT/dt), since the current is proportional to dQ/dt. Since the charge is produced at the surface of the material, due to the change in polarisation, it is therefore not surprising that the magnitude of current is also proportional to the surface area of the electrode on the pyroelectric [52], where

$$i_p = \frac{dQ}{dt} = p \cdot A \cdot \frac{dT}{dt} \tag{6.42}$$

and therefore,

$$i_p = \left[\frac{A}{h}\right] \cdot W \tag{6.43}$$

6.5.1 Performance Figures of Merit for Pyroelectric Harvesting and Sensing

Various pyroelectric-related FoMs are available [52] to select and design materials and composites for both scavenging waste heat and thermal sensing (e.g. thermal imaging, person detectors and fire detection). Figures of merit associated with sensing include those that estimate the current or voltage that is produced due to a thermal source [51, 53, 54].

If we need to optimise the voltage produced from a thermal source (W), we select materials with high voltage responsivity (F_v) and the appropriate FoM [51] is

$$F_v = \frac{p}{c_E \cdot \varepsilon_{33}^{\sigma}} = \frac{p}{\rho \cdot c_p \cdot \varepsilon_{33}^{\sigma}} \tag{6.44}$$

If we aim to optimise the current produced from a thermal source (W), we select materials with a high *current* responsivity (F_i) [51] and the appropriate FoM becomes

$$F_i = \frac{p}{c_E} = \frac{p}{\rho \cdot c_p} \tag{6.45}$$

While the performance FoMs (Eqs. (6.44 and 6.45)) are of interest for thermal sensing, for harvesting a thermal source, we must consider the electrical energy (per cycle thermal, E) and the power ($P = Ef$) that can be produced [50].

In the context of thermal harvesting via the pyroelectric effect [55, 56], the electrothermal coupling factor (k_{pyro}) provides an indication of the efficiency of thermal energy to electrical energy [55]; this is analogous to the electromechanical coupling coefficient (k^2) in Chapter 3, and earlier in this chapter, that provides a measure of the efficiency of mechanical strain energy to electrical energy,

$$k_{pyro}{}^2 = \frac{p^2 \cdot T_{hot}}{c_E \cdot \varepsilon_{33}^{\sigma}} = \frac{p^2 \cdot T_{hot}}{\rho \cdot c_p \cdot \varepsilon_{33}} \tag{6.46}$$

Here, T_{hot} is the upper operating temperature.

A pyroelectric energy harvesting FoM, F_E, is also used to provide an indication of the amount of electrical energy when cycling between two temperatures (T_1 and T_2), rather than simply the efficiency [53], where

$$F_E = \frac{p^2}{\varepsilon_{33}^{\sigma}} \tag{6.47}$$

We can now also develop a pyroelectric thermal harvesting FoM, F_E', based on the electrical energy produced by a thermal source [50]. We will use a similar method to develop FoMs based on the piezoelectric harvesting we used earlier in this chapter [21, 35].

To estimate the stored electrical energy (ΔE) in a pyroelectric harvester due to a change in temperature (ΔT), we assume the pyroelectric material acts as a capacitor where the stored electrical energy is given by $\Delta E = ½ \cdot C \cdot (\Delta V)^2$; therefore,

$$\Delta E = \frac{1}{2} \cdot \left[\frac{p^2}{\varepsilon_{33}^{\sigma}}\right] [A \cdot h] \cdot (\Delta T)^2 \tag{6.48}$$

also,

$$\Delta E = \frac{1}{2} \cdot \left[\frac{p^2}{\varepsilon_{33}^{\sigma} \cdot (c_E)^2}\right] \cdot \left[\frac{A}{h}\right] \cdot (W \cdot \Delta t)^2 \tag{6.49}$$

We can see from the equations above that if the pyroelectric element has a fixed/known area, A, and thickness, h, and the temperature change due to the thermal source is ΔT, and the performance FoM is $p^2/\varepsilon_{33}^{\sigma}$ (Eq. (6.47)); this FoM is relevant when oscillating the pyroelectric materials between two fixed temperatures, for instance with heating and cooling by a pumping cycle of two fluid sources at T_1 and T_2, where $\Delta T = T_2 - T_1$ [57]. However, the thermal harvester may be attempting to scavenge heat from a thermal source with an energy density, W [48]. Based on Eq. (6.49), the performance FoM can therefore be written as

$$F'_E = \frac{p^2}{\varepsilon_{33}^{\sigma} \cdot (c_E)^2} \tag{6.50}$$

This pyroelectric harvesting FoM highlights that optimum pyroelectrics should (i) exhibit a large pyroelectric coefficient, (ii) a low permittivity and (iii) a low volume-specific heat.

We now compare the pyroelectric FoMs, F_v, F_i, k_{pyro}^2, F_E and F'_E for well-known pyroelectric materials; see Table 6.2, which includes FoMs for thermal sensing and thermal harvesting. The volume-specific heat of the materials does not change significantly from material to material; therefore, the current responsivity, F_i, for thermal sensing is large for materials with high pyroelectric coefficients, e.g. polycrystalline PZT and single crystals, such as PMN–PT. Since PVDF has a low relative permittivity, it has a large voltage responsivity,

Table 6.2 Figures of merit F_i, F_v, k_{pyro}^2, F_E and F'_E for various materials, T_{hot} = 298 K for k^2 (%).

Physical parameter	PMN–0.25PT SC[a]	PZT[b]	PVDF[c]	ZnO SC	$LiTaO_3$ SC
p ($\mu C/m^2/K$)	1790 [55]	533 [55]	33 [55]	9.4 [58]	230 [52]
$\varepsilon_{33}^T/\varepsilon_0$	961 [55]	1116 [55]	9 [55]	11 [59]	47 [52]
c_E ($MJ/m^3/K$)	2.5 [55]	2.5 [55]	1.8 [55]	2.3 [60]	3.2 [52]
F_i (pm/V)	716	213	18.3	4.1	71.9
F_v (m^2/C)	0.08	0.02	0.23	0.04	0.17
k_{pyro}^2 (%)	4.5	0.34	0.23	0.01	1.18
F_E ($J/m^3/K^2$)	376.7	28.8	13.6	0.91	127.2
F'_E ($x/10^{12}$)(m^3/J)	60.3	4.6	4.2	0.17	12.4

[a] Relaxor ferroelectric $0.75Pb(Mg_{1/3}Nb_{2/3})O_3$–$0.25PbTiO_3$ single crystal poled along [111] the perovskite unit cell

[b] poled ceramic.

[c] polyvinylidene fluoride, a polymeric ferroelectric.

F_v, for thermal sensing, analogous to the high g_{33} coefficients of the material outlined in Chapter 5 for piezoelectric sensing.

In the context of scavenging waste heat, the $p^2/\varepsilon_{33}^{\sigma}$ energy harvesting FoM for the single crystals and PZT ceramics is large due to their high pyroelectric coefficients. The single crystals have a high F_E' FoM, although the low squared volume-specific heat of PVDF results in a F_E' value that is similar to PZT. Composite materials, for example the porous materials outlined in Chapter 5, could help lead to a large F_E' due to a reduction in both permittivity and volume-specific heat.

The Ashby approach allows a graphical selection of materials [61], and Figure 6.12 is a log–log Ashby plot of pyroelectric performance FoMs, namely the harvesting FOM (F_E), coupling coefficient (k_{pyro}^2) and pyroelectric harvesting FoM (F_E').

As the harvesting FOM, F_E, is on the y-axis, and pyroelectrical material located high on the y-axis has a large F_E; as a result, triglycine sulphate (TGS) and PMN–0.25PT <111> single crystals are good in this context. The non-ferroelectric materials, such as AlN, ZnO, GaN and GdS, have much lower F_E values; remember, it is a log-scale.

It is possible to rearrange the coupling coefficient, k_{pyro}^2 (Eq. (6.46)), as an equation for a straight line ($y = mx + c$), where m is the gradient. Using logs, this leads to

$$\log\left(\frac{p^2}{\epsilon_{33}}\right) = \log(c_E) + \log\left(\frac{k_{pyro}^2}{T_{hot}}\right) \tag{6.51}$$

Equation (6.51) is shown in Figure 6.12 with a gradient of $m = 1$. Materials on the same line of gradient $m = 1$ have the same value of k_{pyro}^2, materials located above the line exhibit a larger k_{pyro}^2. Again, TGS and the single-crystal materials are excellent here.

For the pyroelectric harvesting FoM, F_E', we can also rearrange the equation and take logs to lead to

$$\log\left(\frac{p^2}{\epsilon_{33}}\right) = 2\log(c_E) + \log(F_E') \tag{6.52}$$

This line is shown by a dotted line with a gradient of $m = 2$. Examination of Figure 6.12 reveals that materials with a high F_E' are TGS and the single crystals; PZT and $LiTaO_3$ are also relatively good. As lead-free alternatives, PVDF-based polymers and BNT ceramics have a higher FoM compared to KNN. Again, the non-ferroelectric materials, for example GaN, exhibit relatively low FoMs due to the low pyroelectric coefficients.

Heat transfer and the ability to thermally cycle at a high frequency (f_{max}) have also been recently examined [62], where the maximum pyroelectric power (P_{max}) can be described by

$$P_{max} = Ef_{max} \tag{6.53}$$

For a pyroelectric energy harvester, an FoM was developed, which includes both the electrical energy (E) for a single thermal cycle from T_1 to T_2 and the maximum frequency (f_{max}) by which the harvester can oscillate between the two temperatures while maintaining a homogeneous temperature distribution throughout the material. This is defined as

$$F_{P1} = \frac{p^2}{\epsilon_{33}^{\sigma}}K \tag{6.54}$$

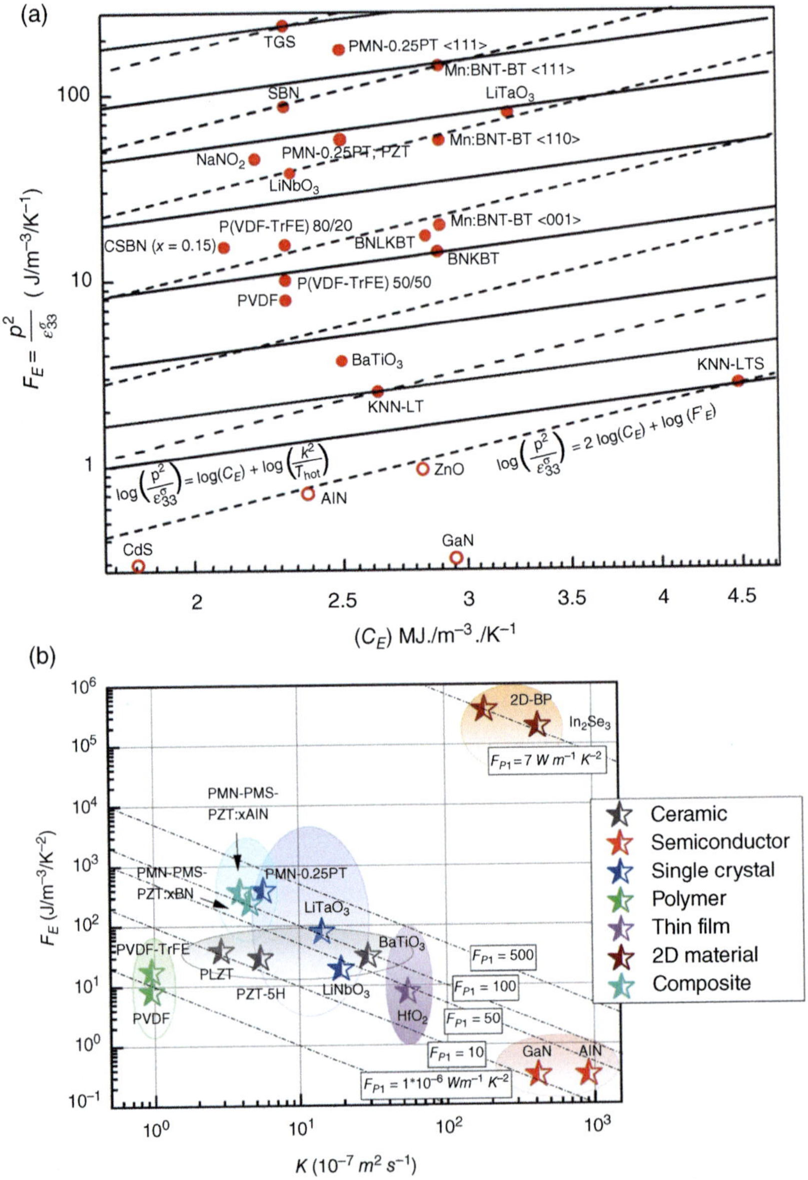

Figure 6.12 (a) Ashby material selection chart for pyroelectric materials. Ferroelectric materials are indicated using solid circles; non-ferroelectric materials are indicated using open circles. *Source:* Bowen et al. [61] / with permission of Royal Society of Chemistry. (b) Ashby chart for selection of materials with high pyroelectric power. *Source:* Narayan et al. [62] / The Royal Society of Chemistry / CC BY 3.0.

where K is the thermal diffusivity of the material. It is clearly advantageous to employ pyroelectric materials that have a high pyroelectric coefficient (to maximise the charge generated as the material cycles between temperatures T_1 and T_2), a low permittivity (to maximise the voltage) and a high thermal diffusivity (to increase the frequency by which the material can be cycled between the temperatures T_1 and T_2). Ashby material selection charts were produced where the two-dimensional pyroelectric materials, In_2Se_3 and 2D black phosphorus, appear attractive due to high pyroelectric coefficients and high thermal diffusivity.

6.6 Summary

This chapter has outlined the range of energy harvesting approaches that use piezoelectric or pyroelectric materials. Mechanical energy harvesting is attracting increasing attention in light of the IoT and Industry 4.0, where there is a need for a wide range of sensors that may require low levels of power or be battery powered. Energy harvesting can extend battery life or even remove the need for batteries completely to provide a fully autonomous system. Mechanical energy harvesting typically employs (i) inertial energy harvesting, which typically operates at a condition of resonance with a spring–mass–damper, and (ii) kinematic energy, where the harvester is coupled to a mechanical base source. Piezoelectric materials, which change their polarisation with stress, are of interest for mechanical harvesting since they enable the solid-state conversion between electrical energy and mechanical energy. Piezoelectric materials also have relatively high energy densities and can therefore be produced at low length scales and can even be integrated into low-power electronic devices such as sensors. Pyroelectric materials, which change polarisation with temperature, are also able to generate electrical energy, but from temperature changes. However, the inherently low frequency of temperature fluctuations certainly provides a limit to the amount of electrical power that can be generated. To select the optimum materials, this chapter has outlined the methods that can be used to select, or even design, materials for energy harvesting. Of particular interest here is the appropriate use of performance FoMs that can be used for selecting materials for mechanical harvesting (stress- and strain-driven applications) and thermal harvesting.

References

1 White, N.M. and Zaghari, B. (2022). Energy harvesting: An overview of techniques for use within the transport industry. *IEEE Electrical Insulation Magazine* 38: 24–32.

2 Guan, M.J. and Liao, W.H. (2008). Characteristics of energy storage devices in piezoelectric energy harvesting systems. *Journal of Intelligent Material Systems and Structures* 19: 671–680.

3 Anton, S.R., Erturk, A., and Inman, D.J. (2010). Multifunctional self-charging structures using piezoceramics and thin-film batteries. *Smart Materials and Structures* 19: 115021.

4 Mane, P., Xie, J., Leang, K.K., and Mossi, K. (2011). Cyclic energy harvesting from pyroelectric materials. *IEEE Transactions on Ultrasonics Ferroelectrics and Frequency Control* 58: 10–17.

5 Bowen, C.R. and Arafa, M.H. (2015). Energy harvesting technologies for tire pressure monitoring systems. *Advanced Energy Materials* 5: 1401787.

6 Kwuimy, C.A.K., Litak, G., Borowiec, M., and Nataraj, C. (2012). Performance of a piezoelectric energy harvester driven by air flow. *Applied Physics Letters* 100: 024103.

7 Dias, J.A.C., De Marqui, C., and Erturk, A. (2013). Hybrid piezoelectric-inductive flow energy harvesting and dimensionless electroaeroelastic analysis for scaling. *Applied Physics Letters* 102: 044101.

8 Beeby, S.P., Tudor, M.J., and White, N.M. (2006). Piezoelectric vibration harvesters based on vibrations of cantilevered bimorphs: a review. *Measurement Science and Technology* 17: R175–R195.

9 Karami, M.A. and Inman, D.J. (2012). Powering pacemakers from heartbeat vibrations using linear and nonlinear energy harvesters. *Applied Physics Letters* 100: 042901.

10 Pytheas Technology. http://pytheas-technology.com (accessed 29 December 2021)

11 Bowen, C.R., Kim, H.A., Weaver, P.A., and Dunn, S. (2014). Piezoelectric and ferroelectric materials and structures for energy harvesting applications. *Energy and Environmental Science* 7: 25–44.

12 van den Ende, D.A., Van de Wiel, H., Groen, W., and van der Zwaag, S. (2012). Direct strain energy harvesting in automobile tires using piezoelectric PZT–polymer composites. *Smart Materials and Structures* 21: 015011.

13 Donelan, J.M., Li, Q., Naing, V. et al. (2008). Biomechanical energy harvesting: generating electricity during walking with minimal user effort. *Science* 319: 807–810.

14 Stewart, M., Weaver, P.M., and M Cain, M. (2012). Charge redistribution in piezoelectric energy harvesters. *Applied Physics Letters* 100: 073901.

15 Roundy, S., Wright, P., and Rabaey, J. (2003). A study of low level vibrations as a power source for wireless sensor nodes. *Computer Communications* 26: 1131–1144.

16 Neri, I., Travasso, F., Mincigrucci, R. et al. (2012). A real vibration database for kinetic energy harvesting application. *Journal of Intelligent Material Systems and Structures* 23: 2095–2101.

17 Kim, S., Priya, S., and Kanno, I. (2012). Piezoelectric MEMS for energy harvesting. *MRS Bulletin* 37: 1039–1050.

18 Zhao, J., Zhen, X., Zhou, L. et al. (2012). Investigation of a d_{15} mode PZT51 piezoelectric energy harvester with a series connection structure. *Smart Materials and Structures* 21: 105006.

19 Erturk, A. and Inman, D.J. (2011). *Piezoelectric Energy Harvesting*. Wiley.

20 Cho, J., Richards, R., Bahr, D. et al. (2006). Efficiency of energy conversion by piezoelectrics. *Applied Physics Letters* 89: 104107.

21 Roscow, J.I., Pearce, H., Khanbareh, H. et al. (2019). Modified energy harvesting figures of merit for stress-and strain-driven piezoelectric system. *The European Physical Journal Special Topics* 228: 1537–1554.

22 Uchino, K. (2018). Piezoelectric energy harvesting systems—essentials to successful developments. *Energy Technology* 6: 829–848.

23 Deutz, D.B., Pascoe, J.-A., Schelen, B. et al. (2018). Analysis and experimental validation of the figure of merit for piezoelectric energy harvesters. *Materials Horizons* 5: 444–453.

24 Islam, R.A. and Priya, S. (2006). Realization of high-energy density polycrystalline piezoelectric ceramics. *Applied Physics Letters* 88: 032903.

25 Crossley, S. and Kar-Narayan, S. (2015). Energy harvesting performance of piezoelectric ceramic and polymer nanowires. *Nanotechnology* 26: 344001.

26 Rödig, T., Schönecker, A., and Gerlach, G. (2010). A Survey on piezoelectric ceramics for generator applications. *Journal of the American Ceramic Society* 93: 901–912.

27 Uchino, K. and Giniewicz, J.R. (2003). *Micromechatronics*. New York: CRC Press.

28 Xiao, H., Pan, M., Harry Chu, J.Y. et al. (2022). Hydraulic pressure ripple energy harvesting: structures, materials, and applications. *Advanced Energy Materials* 12: 2103185.

29 Cunefare, K.A., Skow, E.A., Erturk, A. et al. (2013). Energy harvesting from hydraulic pressure fluctuations. *Smart Materials and Structures* 22: 025036.

30 St. Clair, D., Bibo, A., Sennakesavababu, V.R. et al. (2010). A scalable concept for micropower generation using flow-induced self-excited oscillations. *Applied Physics Letters* 96: 144103.

31 Verma, N., Skow, E. A., Cunefare, K. A., Erturk, A. (2013) *Proceedings 2013 Conference on Smart Materials*, Adaptive Structures and Intelligent Systems, ASME, Snowbird.

32 Skow, E.A., Cunefare, K.A., and Erturk, A. (2014). Power performance improvements for high pressure ripple energy harvesting. *Smart Materials and Structures* 23: 104011.

33 Skow, E. A., Cunefare, K. A., Erturk, A. (2014). *Proceedings 2014 International Mechanical Engineering Congress and Exposition*, ASME, Canada 2014, 4.

34 Xu, T. B., Siochi, E., Kang, J. et al. (2011) *Proceedings ASME Design Engineering Technical Conference*, Washington.

35 Priya, S. (2010). Criterion for material selection in design of bulk piezoelectric energy harvester. *IEEE Transactions on Ultrasonics Ferroelectrics and Frequency Control* 57: 2610–2612.

36 Yang, Z.B., Zhou, S.X., Zu, J., and Inman, D. (2018). High-performance piezoelectric energy harvesters and their applications. *Joule* 2: 642.

37 Skow, E. A. (2017). *Harvesting energy from acoustic pressure fluctuations within hydraulic systems via excitation of piezoelectric stacks*, PhD Thesis. Georgia Institute of Technology.

38 Zhang, S.J. and Li, F. (2012). High performance ferroelectric relaxor-PbTiO3 single crystals: status and perspective. *Journal of Applied Physics* 111: 031301.

39 Belincout, D., Kruger, H. H. A. (2001) Properties of Morgan Electroceramics, Technical Publication TP-226.

40 TRS Technologies http://www.trstechnologies.com/Materials/High-Performance-PMN-PT-Piezoelectric-Single-Crystal (accessed 29 December 2021)

41 Cao, H. and Evans, A.G. (1993). Nonlinear deformation of ferroelectric ceramics. *Journal of the American Ceramic Society* 76: 890.

42 Hooker, M. W. (1998). Properties of PZT-based piezoelectric ceramics between -150 and 250 degC. NASA/CR-1998-208708. 1998.

43 Damjanovic, D. (1998). Materials for high temperature piezoelectric transducers. *Current Opinion in Solid State and Materials Science* 3: 469.

44 Aranda, J. J. L, Oelmann, B., Bader, S. (2017). *IEEE International Conference on Advanced Intelligent Mechatronics (AIM)*, Munich 2017.

45 Cao, D.X., Duan, X.J., Guo, X.Y., and Lai, S.K. (2020). Design and performance enhancement of a force-amplified piezoelectric stack energy harvester under pressure fluctuations in hydraulic pipeline systems. *Sensors and Actuators A* 309: 112031.

46 Xiao, H.F., Qie, H., and Bowen, C.R. (2021). Modelling of the circular edge-clamped interface of a hydraulic pressure energy harvester to determine power, efficiency and bandwidth. *Mechanical Systems and Signal Processing* 146: 107013.

47 Zhang, Q., Agbossou, A., Feng, Z. et al. (2011). Solar micro-energy harvesting with pyroelectric effect and wind flow. *Sensors and Actuators A* 168: 335–342.

48 Krishnan, S., Ezhilarasi, D., Uma, G., and Umapathy, M. (2014). Pyroelectric-based solar and wind energy harvesting system. *IEEE Transactions on Sustainable Energy* 5: 73–81.

49 Kotipalli, V., Gong, Z., Pathak, P. et al. (2010). Light and thermal energy cell based on carbon nanotube films. *Applied Physics Letters* 97: 124102.

50 Bowen, C.R., Topolov, V.Y., Taylor, J. et al. (2015). A modified figure of merit for pyroelectric energy harvesting. *Materials Letters* 138: 243–246.

51 Lang, S.B. and Das-Gupta, D.K. (2001). Pyroelectricity: fundamentals and applications. In: *Handbook of Advanced Electronic and Photonic Materials and Devices*, vol. 4 (ed. H.S. Nalwa), 1–54. Academic Press.

52 Whatmore, R.W. (1986). Pyroelectric devices and materials. *Reports on Progress in Physics* 49: 335–1386.

53 Cooper, J. (1962). A fast response pyroelectric thermal detector. *Journal of Scientific Instrumentation* 39: 467–472.

54 Putley, E.H. (1980). A method for evaluating the performance of pyroelectric detectors. *Infrared Physics* 20: 139–146.

55 Sebald, G., Lefeuvre, E., and Guyomar, D. (2008). Pyroelectric energy conversion: optimisation principles. *IEEE Transactions on Ultrasonics Ferroelectrics and Frequency Control* 55: 538–551.

56 Sebald, G., Seveyrat, L., Guyomar, D. et al. (2006). Electrocaloric and pyroelectric properties of 0.75Pb(Mg1/3 Nb2/3)O3 -0.25PbTiO3 single crystals. *Journal of Applied Physics* 100: 24112.

57 Ravindran, S.K.T., Huesgen, T., Kroener, M., and Woias, P. (2011). A self-sustaining micro thermomechanic-pyroelectric generator. *Applied Physics Letters* 99: 104102.

58 Lang, S.B. (2005). Pyroelectricity: From ancient curiosity to modern imaging tool. *Physics Today* 58: 31–36.

59 Young, K.F. and Frederikse, H.P.R. (1973). Compilation of the static dielectric constant of inorganic solids. *Journal of Physical and Chemical Reference Data* 2: 313–409.

60 El-Brolossy, T.A., Saber, O., and Ibrahim, S.S. (2013). Determining the thermophysical properties of Al-doped ZnO nanoparticles by the photoacoustic technique. *Chinese Physics B* 22: 07440.

61 Bowen, C.R., Taylor, J., LeBoulbar, E. et al. (2014). Pyroelectric materials and devices for energy harvesting applications. *Energy and Environmental Science* 7: 3836–3856.

62 Narayan, B., Wang, Q., Roscow, J. et al. (2025). New pyroelectric figures of merit for harvesting dynamic temperature fluctuations. *Journal of Materials Chemistry A* 13: 12977–12987.

7

Drive Electronics and Control

7.1 Introduction

For open-loop control of a piezoelectric actuator, a power electronic drive amplifier is required to convert a low-voltage, low-current control signal into a high-voltage signal with sufficient current capability to deliver the required power to the actuator. In addition, for more accurate actuator position control, output position measurement is desirable, which can be fed back to a controller to enable closed-loop position control.

Piezoelectric actuators require power amplifiers with a fast response and high output voltage. These circuits are usually implemented using power operational amplifiers (op-amps). The amplifier must be designed (or specified) to achieve the required voltage and current, which will determine the rate of change in voltage and hence the maximum velocity for an actuator of given capacitance. Although there is very little holding current because the actuator is essentially a capacitive device, power levels can be high if the actuator is switched at a high repetition rate. For example, an actuator of 8.0 μF capacitance, with a 100 Hz 500 V driving voltage, requires an input power of 100 W. Note that although not common, a regenerative circuit is possible whereby up to 90% of the energy can be regained upon discharge of the actuator.

Fundamentally, there are two options to drive a piezoelectric actuator: voltage and charge amplifiers. Voltage amplifiers are the norm since the circuit design is simple and reliable. Its major drawback is that the displacement/voltage curve will exhibit hysteresis and creep phenomena. With a charge amplifier, these effects are virtually absent: there is an almost linear relation between dielectric displacement and deformation. However, charge amplifier circuitry is more complicated, and actuator displacement is more sensitive to temperature changes in this drive mode.

The focus of this chapter is piezoelectric actuators; however, it should be noted that piezoelectric sensors also require appropriate signal conditioning and amplification, i.e. a conversion of the charge output into a voltage of a useable magnitude. This is discussed in Section 7.8.

7.2 Op-Amp Circuits

Piezoelectric power amplifiers usually employ op-amp circuits. These circuits all include feedback, i.e. a connection back from the op-amp output to one or the other of the inputs.

Piezoelectric Materials: Properties, Applications and Devices, First Edition.
Christopher R. Bowen and Andrew R. Plummer.

Figure 7.1 shows two simple examples of op-amp circuits. To a good approximation, op-amps with feedback have the following characteristics:

i) the voltages e_1 and e_2 at the two inputs are the same – i.e. the output of the op-amp will vary in whatever way is required in order to keep the inputs at the same voltage. This is a result of an extremely high gain between the voltage difference at the input and the output voltages of the op-amp. Note that in practice, there will be a maximum output voltage dictated by the op-amp power supply.
ii) the inputs draw no current – i.e. the op-amp has an extremely high input impedance.
iii) any current (up to some limit) can be drawn from the output without altering the output voltage – i.e. the op-amp has an extremely low output impedance.

These rules can be used to determine the basic behaviour of an op-amp circuit. The purpose of the voltage amplifier circuits shown in Figure 7.1 is to provide a scaling factor between the input and the output voltages and to ensure that the output voltage is insensitive to the current drawn from the output terminal by the attached load.

We consider the non-inverting amplifier as an example. The high gain of the op-amp ensures $e_2 = e_1 = V_i$. The current is the same through resistors R_1 and R_2 and is given by:

$$\frac{V_i}{R_1} = \frac{V_o - V_i}{R_2} \tag{7.1}$$

which can be rearranged to give the voltage gain:

$$V_o = \left(1 + \frac{R_2}{R_1}\right) V_i \tag{7.2}$$

Other circuits can be analysed in a similar way. The following sections discuss circuit design in much greater detail, including practical examples and limitations relevant to

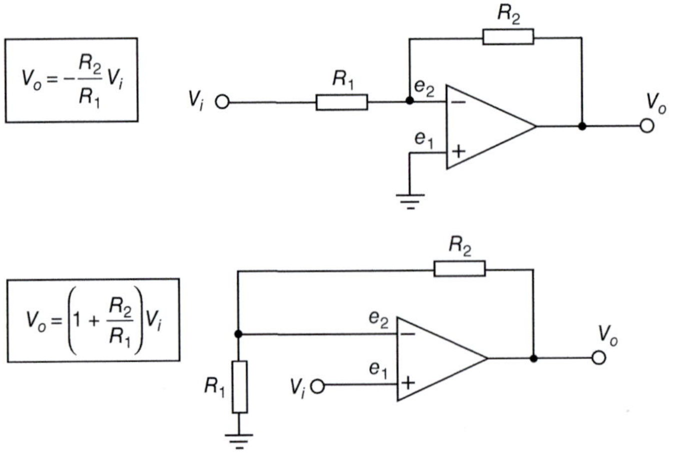

Figure 7.1 Two common op-amp circuits: an inverting voltage amplifier (top) and a non-inverting voltage amplifier (bottom).

piezoelectric power amplifiers. Note that these are for using piezoelectric ceramics as actuators, rather than the type of signal amplifier required for use in sensing.

7.3 Voltage Amplifiers for Driving Actuators

In the first example, the circuit shown in Figure 7.2 is designed to drive a piezoelectric actuator of 0.5 μF capacitance. The circuit provides a maximum output slew rate of 20 V/μs, which will dictate the maximum actuator velocity, and also requires a current of 10 A when connected to the 0.5-μF actuator. To prevent a current overload, it is important that the slew rate is limited to the given value. This is achieved with the compensation networks formed by capacitors C_3 and C_4 and resistors R_3 and R_4, as defined by the manufacturer's data sheet for the particular op-amp, in this case, the PA69. As the PA69 is unable to deliver the required 10 A, it drives a MOSFET pair, X_1 and X_2, in order to deliver the current.

The circuit is a non-inverting amplifier configuration. Substituting the values of resistors R_1 and R_2 into Eq. 7.2 leads to a gain of 10. Thus, the input voltage range of ±4.8 V is amplified up to the output voltage range of ±48 V, which is the maximum voltage range dictated by the op-amp power supply voltage.

Resistors R_5 and R_6 are the gate resistors for the MOSFETs that dissipate energy to prevent high-frequency oscillations. Resistor R_7 modifies the output load so that it is not completely capacitive, improving stability.

Figure 7.3 shows a circuit for a higher voltage driver, capable of applying ±150 V across the actuator. The left op-amp (A) is an inverting amplifier connected to one side of the actuator (V_A), and the right op-amp (B) is an inverting amplifier connected to the other actuator terminal (V_B). With zero input voltage V_1, both V_A and V_B will sit at 85 V, giving zero field in

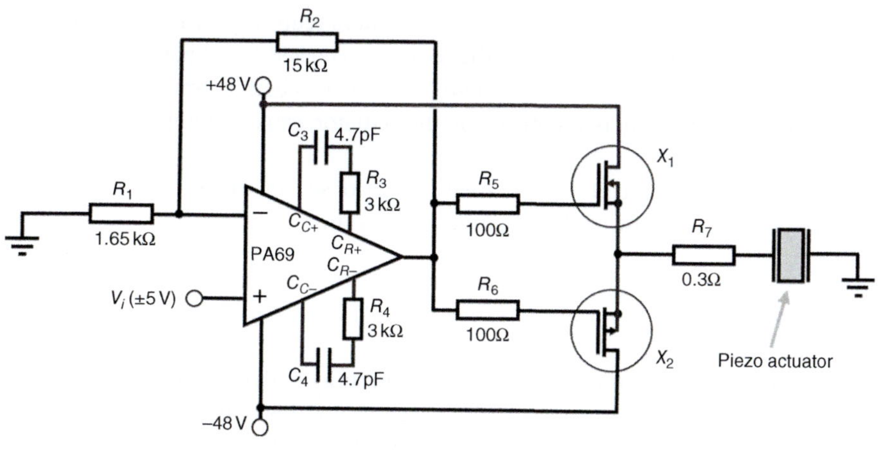

Figure 7.2 Driver circuit delivering ±48 V and slew rates of 20 V/μs. Adapted from [1].

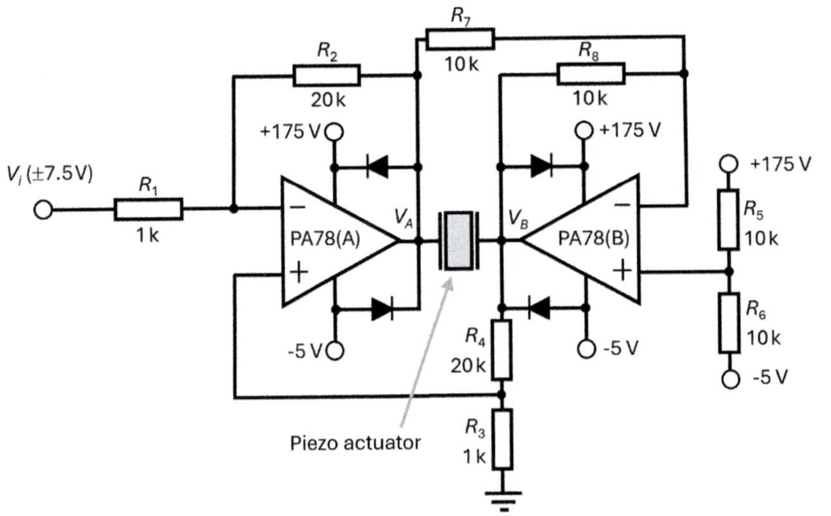

Figure 7.3 Example ±150-V piezo-amplifier, using two PA78 op-amps. Adapted from [1].

the actuator. This voltage is fixed by the voltage divider formed by R_5 and R_6; both inputs to op-amp B will be at 85 V in this condition, and the R_4/R_3 ratio means that the non-inverting input to op-amp A will be at 4 V. As this ratio is the same as the gain R_2/R_1, the inverting input to op-amp A will also be at 4 V as expected.

With the input voltage V_i varying in the range of −7.5 V to +7.5 V, V_A swings down by 150 V, from 160 V to 10 V, the gain being less than the R_2/R_1 expected for a simple inverting amplifier due to the varying bias voltage on the non-inverting input. Op-amp B is acting as an inverting amplifier with gain given by R_8/R_7, which is unity, and input V_A, so the corresponding change in its output V_B is a positive swing of 150 V, from 10 V to 160 V.

Other components (not shown) are used to limit the current and voltage slew rate. Note that this circuit is designed for small capacitance actuators requiring currents within the maximum current capability of the op-amps themselves, so no additional power transistors are required. The four diodes connect the actuator terminals to both power supply rails to protect the op-amps in the event of an impact on the actuator, causing a high piezoelectric voltage to be generated.

The two voltage amplifier circuits shown are examples of several different configuration options. These options are illustrated in Figure 7.4. They are:

a) *Direct op-amp control*: Suitable for low-power applications only, as the output current of the high-voltage op-amps is typically less than 200 mA. The circuit in Figure 7.3 is based on this configuration.
b) *Power boost follower*: With a transistor power stage in series with the feedback amplifier. (The circuit in Figure 7.2 is of this type.) The accuracy of this amplifier is reduced, as there is no feedback directly from the voltage applied to the piezoelectric actuator.
c) *Error power boost type*: The output voltage of the power boost stage is directly fed back to the op-amp circuit.

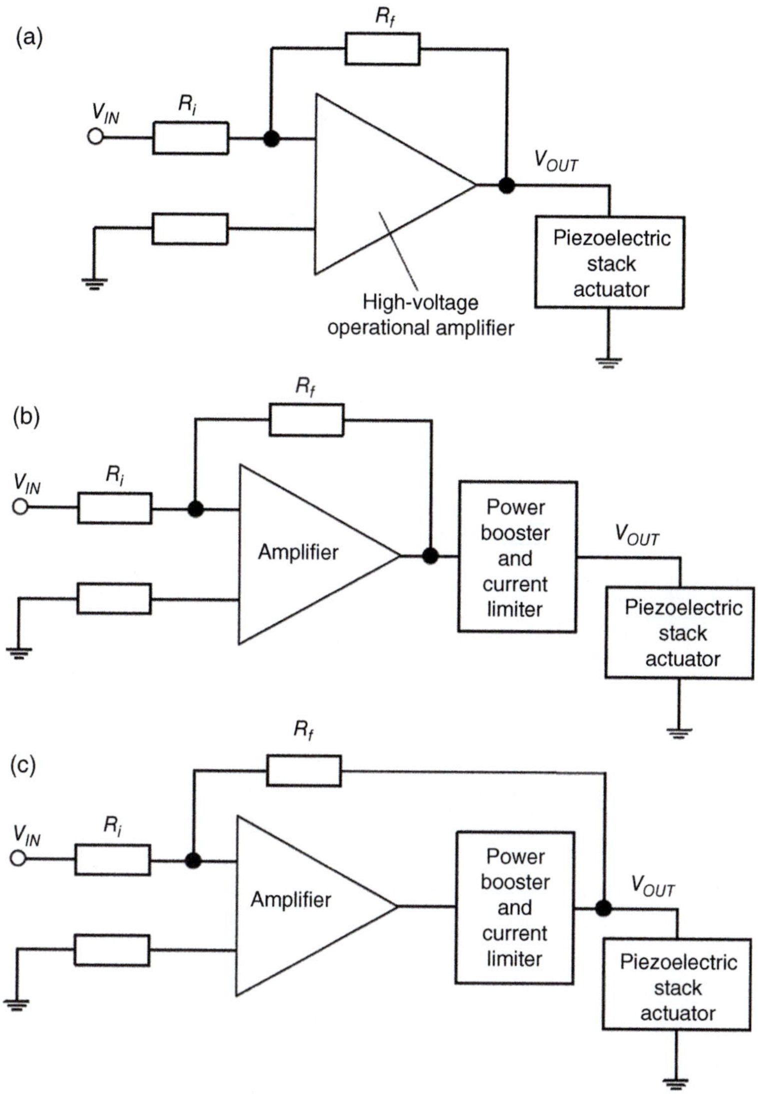

Figure 7.4 Voltage amplifier configurations: (a) direct op-amp control, (b) power boost follower and (c) error power boost type.

7.4 Charge Amplifiers for Driving Actuators

As shown in Figure 7.5, although the relationship between voltage and displacement exhibits considerable hysteresis for a typical piezoelectric actuator, the charge-to-displacement characteristic is much more linear. Thus, driving an actuator using a charge amplifier rather than a voltage amplifier can reduce the hysteresis by an order of magnitude.

Figure 7.6 shows a simple charge amplifier, based on an inverting op-amp circuit and a sensing capacitor C_s. The principle is that the current through the sensing capacitor is

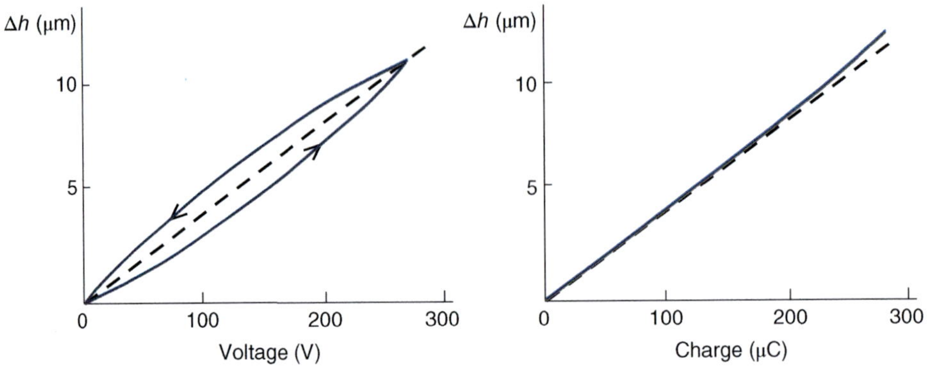

Figure 7.5 Displacement (Δh) versus voltage (left) and charge (right) for an example piezoelectric actuator. Adapted from [2].

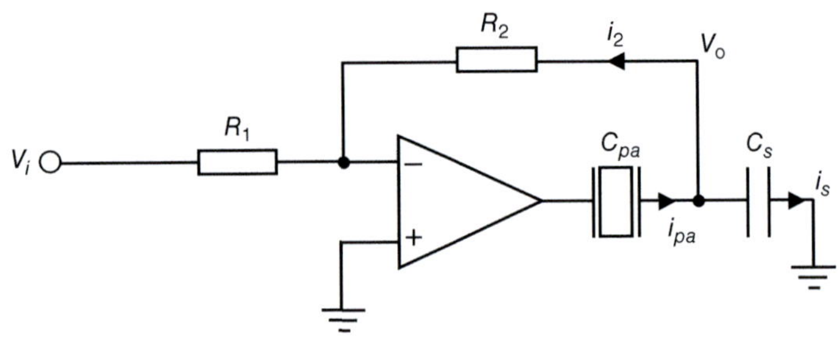

Figure 7.6 Charge amplifier with no drift compensation.

approximately the same as that through the piezoelectric actuator, i.e. $i_{pa} \approx i_s$, and if the latter acts as a capacitor, the charge developed (q_{pa}) will be the same as that in C_s:

$$q_{pa} \approx q_s = C_s V_o$$

$$q_{pa} \approx -C_s \frac{R_2}{R_1} V_i \tag{7.3}$$

In reality, there will be a small current through the feedback resistor R_2, so:

$$i_{pa} = i_s + i_2$$

$$i_{pa} = C_s \frac{dV_o}{dt} + \frac{V_o}{R_2} \tag{7.4}$$

Integrating Eq. (7.4):

$$q_{pa} = -C_s \frac{R_2}{R_1} V_i + \frac{1}{R_2} \int V_o \, dt \tag{7.5}$$

The second term in Eq. (7.5) means that the actuator charge will drift, as the output voltage will inevitably not have a mean of exactly zero, and the rate of drift can be reduced but never eliminated by making R_2 large. Thus, the op-amp output will drift until it saturates at its power supply voltage level.

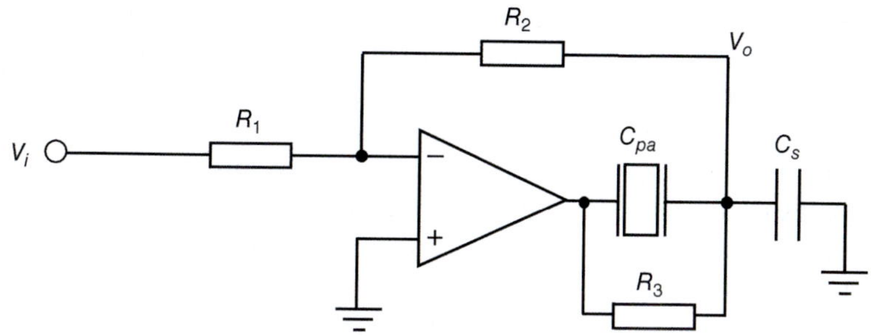

Figure 7.7 Charge amplifier with drift compensation resistor (R_3).

In Figure 7.7, an additional resistor has been placed in parallel with the piezoelectric actuator to tackle this problem. Now, the sum of the currents at the output of the actuator gives an extra term compared to Eq. (7.4):

$$i_{pa} = C_s \frac{dV_o}{dt} + \frac{V_o}{R_2} - \frac{q_{pa}}{C_{pa}R_3}$$

$$\frac{dq_{pa}}{dt} + \frac{q_{pa}}{C_{pa}R_3} = C_s\left(\frac{dV_o}{dt} + \frac{V_o}{C_sR_2}\right) \tag{7.6}$$

If the resistor R_3 is chosen such that:

$$R_3 = \frac{C_sR_2}{C_{pa}} \tag{7.7}$$

Then, Eq. (7.6) reduces to:

$$q_{pa} = C_sV_o$$

$$q_{pa} = -C_s\frac{R_2}{R_1}V_i \tag{7.8}$$

and we can control the actuator charge proportionally without drift.

In practice, as the piezoelectric actuator does not behave perfectly like a capacitor with constant C_{pa}, the value for R_3 can only be an approximation. This leads to an error in the steady-state gain of the amplifier, although this is only significant up to the break frequency (the reciprocal of the time constant) present in Eq. (7.6), which is:

$$\frac{1}{C_{pa}R_3}\ \text{rad/s} \tag{7.9}$$

In a commercial charge amplifier, R_3 is often fixed, so inaccuracy at low frequency is expected, and the amplifier has a specified frequency range, which starts above the break frequency given by Eq. (7.9). For example, for a 1-μF actuator, the amplifier may have a minimum operation frequency of 0.1 Hz.

Due to this problem, other methods for tackling drift have also been proposed; for example, including an extra feedback loop with an integrating amplifier to drive down the output voltage of the op-amp [3]. If we return to the simple amplifier in Figure 7.6, it may be tempting to put a buffer in the V_o feedback line so that the current i_2 does not need to be drawn from the actuator–capacitor junction. However, the drift problem will still remain due to actuator leakage current and other non-capacitor-like characteristics, which means the voltage across the actuator will diverge from that across the sensing capacitor (scaled by the capacitance ratio). It should be noted that if the charge amplifier is being used within another closed-loop control system, for example, using actuator position feedback, then the drift may not be a problem, as it will be compensated by this outer loop controller.

Another problem, inherent to all charge amplifiers, is temperature sensitivity. The permittivity of piezoceramic changes with temperature, so, for example, the strain per charge density (g_{ij} coefficient) may decrease by 10% as temperature changes from 0 °C to 100 °C, whereas the displacement per voltage (d_{ij} coefficient) is nearly constant in the same temperature range (see Section 5.4) [4]. Thus, the charge amplifier gain, in terms of input voltage to actuator position, is quite variable with temperature for common piezoceramics.

Note 7.1

A piezoelectric actuator driven by a voltage amplifier exhibits significant hysteresis. A charge actuator gives better linearity, but its gain is sensitive to temperature, and low-frequency drift is difficult to eradicate.

7.5 Regenerative Amplifiers

As the piezoelectric actuator behaves largely as a capacitor, a significant fraction of the electrical charging power is reactive – it is fed back into the amplifier output during the discharge cycle. Linear amplifiers convert this to power heat internally, so that the overall efficiency is low. Switching amplifiers can be designed for energy recovery: they allow the discharge energy to be driven back into the power rails.

Figure 7.8 shows a switching amplifier. A pulse width modulation (PWM) drive voltage V_{PWM} is converted to a switching signal A+ as input to one transistor, and the mapping

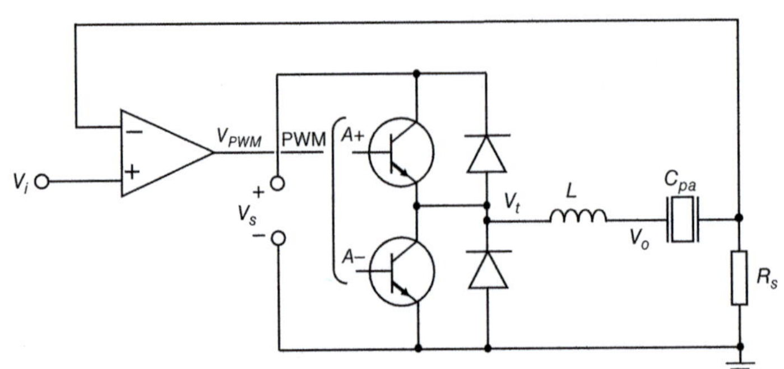

Figure 7.8 Switching amplifier with regenerative capability.

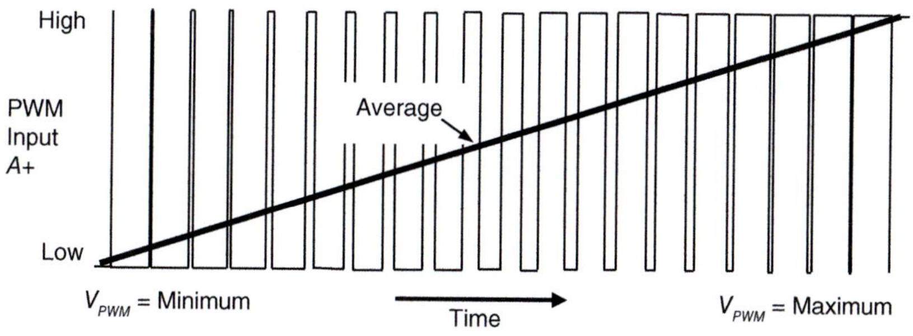

Figure 7.9 Pulse width modulation (PWM) signal.

between voltage and mark–space ratio is indicated in Figure 7.9. The second transistor is switched according to A–, which is the inverse of A+. The consequence is that voltage V_t is rapidly switched between the power rails, such that its average value is proportional to V_{PWM}. A pair of diodes limits V_t to remain within the voltage range of the power supply. The switching frequency is high, e.g. 100 kHz.

An inductor smoothes the flow of current into the actuator. In the circuit shown, the voltage from a current sensing resistor R_s is fed back to the op-amp, so in fact, this is a current amplifier. A second op-amp with feedback from voltage V_o could be added to convert this into a voltage amplifier; as long as the voltage drop across R_s is small, V_o is a good approximation to the voltage across the actuator.

The current through the actuator can be determined by summing the voltages across the inductor, actuator and sensing resistor:

$$V_t = L\frac{di}{dt} + \frac{q_{pa}}{C_{pa}} + iR_s \tag{7.10}$$

$$\frac{dV_t}{dt} = L\left(\frac{d^2i}{dt^2} + \frac{R_s}{L}\frac{di}{dt} + \frac{i}{LC_{pa}}\right) \tag{7.11}$$

This indicates that the response of the current i is a second-order lag in relation to the rate of change in voltage V_t, with natural frequency:

$$\frac{1}{\sqrt{LC_{pa}}} \text{ rad/s} \tag{7.12}$$

If this natural frequency is chosen to be well below the switching frequency, say 1 kHz (6280 rad/s), then the current will just respond to the mean value of V_t within each PWM cycle, and Eq. (7.11) becomes:

$$\left.\frac{dV_t}{dt}\right|_{mean} \approx \frac{i}{C_{pa}} \tag{7.13}$$

As a consequence, the actuator charge is proportional to the voltage V_{PWM}, and the overall behaviour of the switching amplifier is very similar to a linear amplifier, but with much greater efficiency. More details of this type of amplifier and variations thereof can be found in [5, 6].

Note 7.2

Linear voltage amplifiers are very commonly used to drive piezoelectric actuators. Regenerative amplifiers are more complex and costly, but the result is significantly improved energy efficiency.

7.6 Position Sensors for Feedback Control

The displacement of a piezoelectric actuator is, to a first approximation, proportional to the applied voltage. However, accurate control of displacement open loop by adjusting voltage is hampered by:

i) movement under load,
ii) temperature sensitivity and
iii) piezoelectric hysteresis (reduced by using a charge amplifier as described in Section 7.3, but this introduces other problems, notably poor low-frequency control).

Thus, using a position sensor and closed-loop control will improve accuracy. This is illustrated in Figure 7.10. There are several options for accurately sensing the position over the relatively small strokes involved.

7.6.1 Linear Variable Differential Transformer (LVDT)

In an LVDT, the displacement of an iron core (sometimes called a slug) is measured by the alteration of inductance between a primary winding and two secondary windings. As shown in Figure 7.11, the primary winding carries an alternating current of constant amplitude.

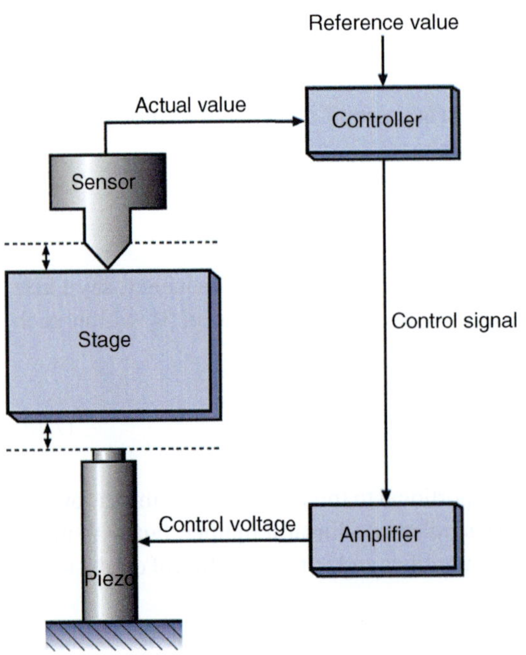

Figure 7.10 Closed-loop position control.

Figure 7.11 LVDT with core in mid-position (top) and displaced (bottom).

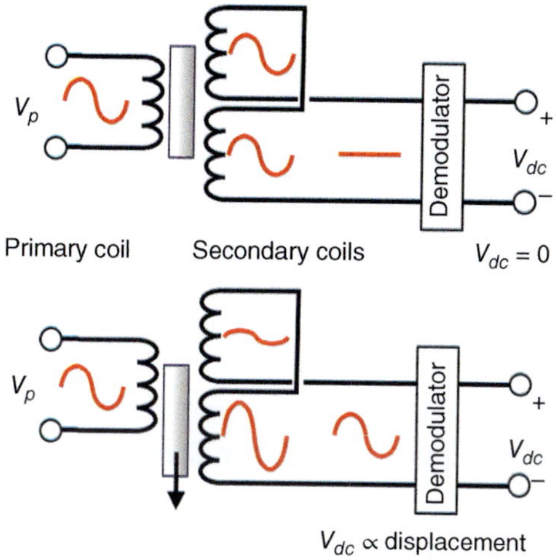

With the iron core in the mid-position, the induced voltages in the two secondary windings are equal, but the output voltage is zero, as it is the difference between these two voltages. With the iron core displaced, one secondary winding has a greater induced voltage than the other, and so, the output voltage is non-zero. The demodulator converts the alternating output voltage into a constant (DC) voltage proportional to its amplitude. The LVDTs are compact, wear-free and relatively cheap. They are most suited to measuring the displacement of bending actuators or mechanically amplified stacks (displacements of ~0.2 mm or more).

7.6.2 Eddy Current Sensor

A coil is supplied with an alternating current producing an electromagnetic field; see Figure 7.12. If an electrically conducting object is placed in this field, eddy currents are induced, which generate an electromagnetic field that opposes the coil field and changes

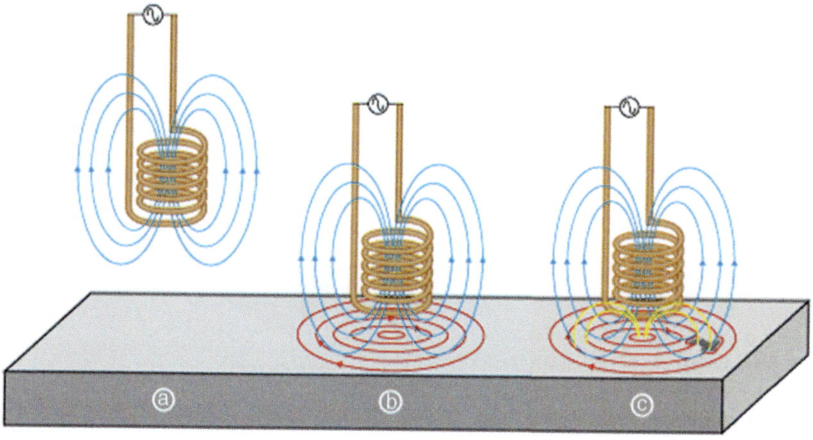

Figure 7.12 Eddy current sensor.

its impedance. A controller measures the change in impedance and derives distance. Eddy current sensors are very rugged and accurate, but the controllers are relatively expensive.

7.6.3 Capacitive Sensor

The sensor and target operate like the electrodes in a parallel plate capacitor; see Figure 7.13. If an AC with constant frequency flows through the sensor, the voltage is proportional to the distance to the target. Capacitive sensors are sensitive to contamination in the sensor gap and have limited cable length but possess good linearity, resolution and temperature stability.

7.6.4 Laser Triangulation Sensor

Laser triangulation sensors determine the position of a target by measuring light reflected from the target surface; see Figure 7.14. A transmitter (laser diode) projects a spot of light onto the target, and its reflection is focused via an optical lens onto a light-sensitive receiver. If the target changes its position, the position of the reflected spot of light on the detector changes as well. These sensors can be relatively accurate but sensitive to target surface conditions and tilt.

7.6.5 Strain Gauge Sensor

The *bonded metal foil* strain gauge is the most common type and consists of thin copper or aluminium foil cut in the pattern shown in Figure 7.15 and bonded to a non-conducting base. The base is then glued to the piezoelectric stack or bender where the strain is to be measured. Note that this does not give a direct measurement of actuator end-position – if the deflection of the beam shown in the figure is due to both piezoelectric strain and external force, then strain gauges at two positions along the beam are required to estimate the end deflection.

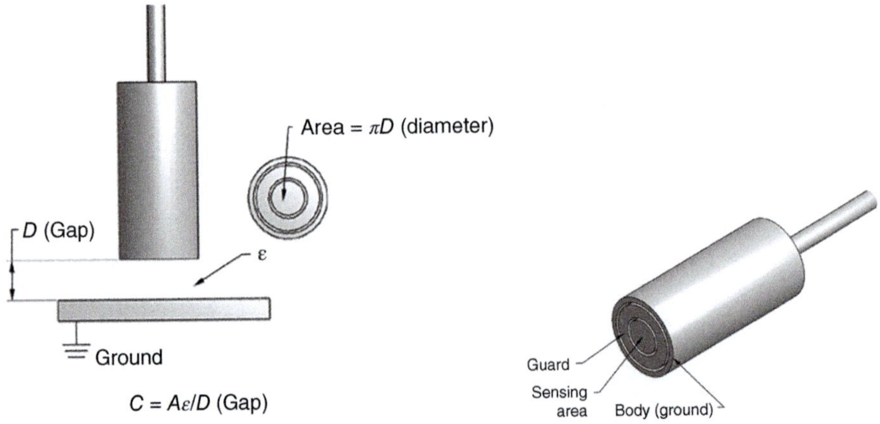

Figure 7.13 Capacitive sensor.

Figure 7.14 Laser triangulation sensor.

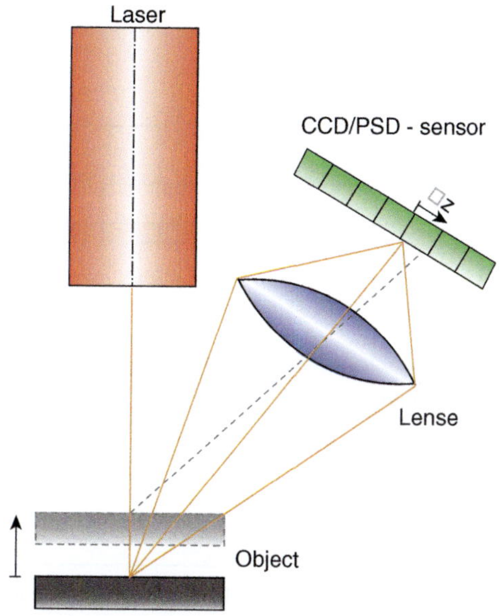

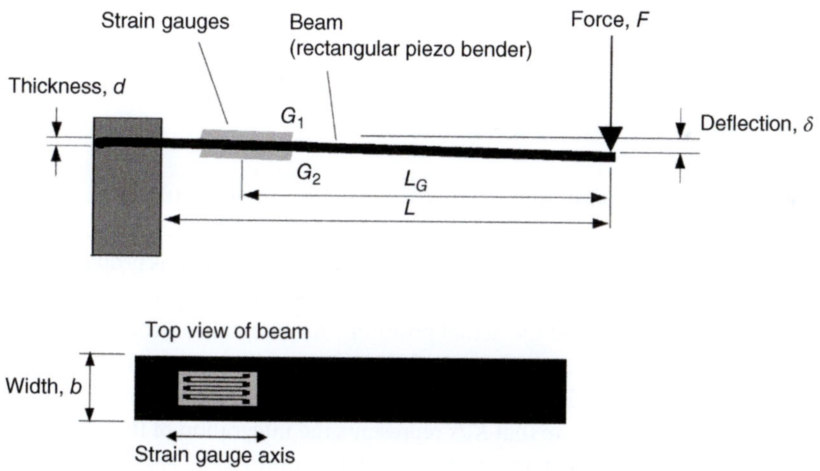

Figure 7.15 Strain gauges bonded to a rectangular piezoelectric bender.

If the gauge is stretched along its axis, its resistance increases in proportion to the strain; if it is compressed, its resistance decreases. The change in resistance of the gauges is measured using a Wheatstone bridge and differential amplifier, as shown in Figure 7.16 for the arrangement of Figure 7.15. As the two strain gauges are on opposite sides of the beam, one increases in resistance, while the other decreases, and the bridge arrangement shown is suitable. The advantages of using two gauges as opposed to one are double the output and less temperature sensitivity. Strain gauges are of low cost but can be sensitive to electrical noise.

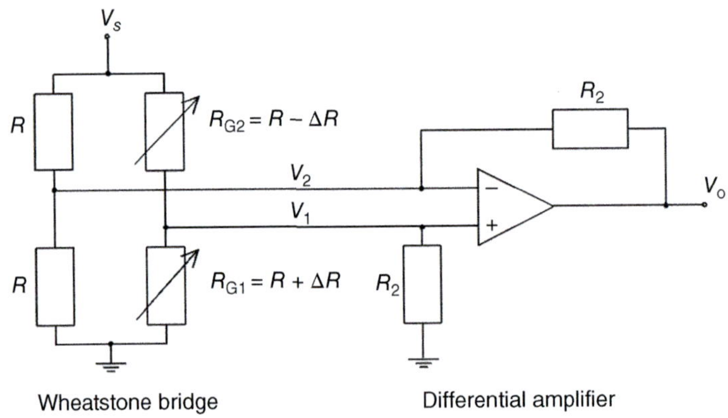

Figure 7.16 Strain gauge bridge and amplifier circuit.

> **Note 7.3**
>
> Feedback control using an accurate position sensor is required for precise piezoelectric actuator positioning. Open-loop control, relying on the piezoelectric voltage–displacement relationship, is subject to errors due to compliance (i.e. movement under external loading), temperature sensitivity, creep and hysteresis.

7.7 Closed-Loop Controllers: Case Study

Figure 7.17 shows a closed-loop position controller suitable for a piezoelectric actuator. As the plant characteristic is one in which the position is roughly proportional to the amplifier input signal, just using a proportional controller in which the amplifier input is proportional to the position error ($K_i = 0$, $K_{ff} = 0$ in the controller shown) results in a significant steady-state error between the demanded and the actual positions. An integral controller is better ($K_p = 0$, $K_{ff} = 0$) in which the amplifier input is the error integrated over time and will give zero steady-state error, as the amplifier input signal can be non-zero even when actual and demanded positions are the same. Note that K_i/s represents the integration of the error signal and scaling by the gain K_i. The dynamic response can be improved by also including a proportional term (non-zero K_p), but an alternative and effective way of improving the dynamic response is to include a demand feed-forward path in which K_{ff} is calculated to give the correct amplifier input signal for the position demand based on the approximate proportionality between voltage and position for the piezoelectric actuator. The gain K_{ff} may need to be reduced below this ideal value to prevent 'overdriving' the actuator during transients, resulting in overshoot. Note that including integral anti-wind-up is also important to prevent the integral term from increasing indefinitely in circumstances where the error cannot be driven to zero.

Figure 7.18 shows the closed-loop response of an actuator using the controller of Figure 7.17. The actuator is a 40-mm-diameter ring bender, with a free displacement of about ±100 μm and a blocking force of ±40 N, at a maximum drive voltage of ±100 V. In

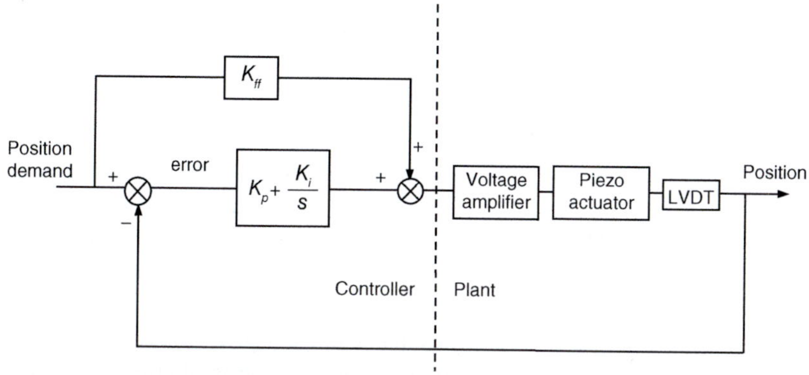

Figure 7.17 Proportional–integral (PI) controller with demand feed-forward.

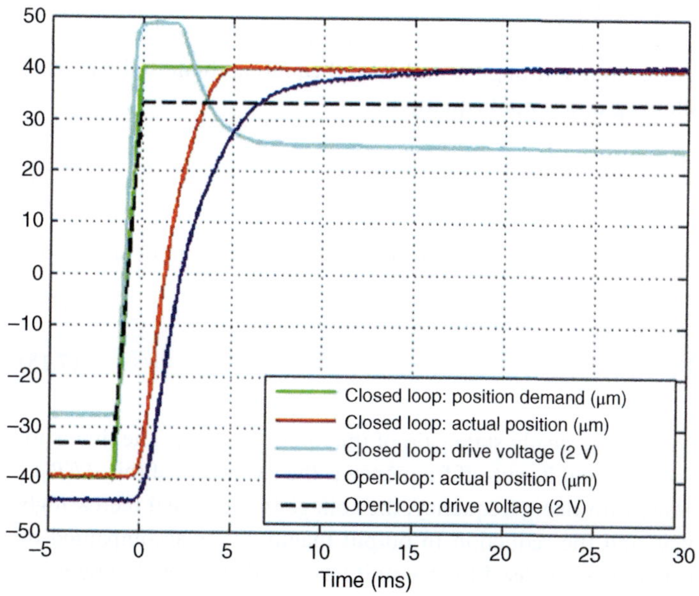

Figure 7.18 Closed-loop ring bender position response compared to open-loop response [7] / with permission of ASME.

this case, it is operating against a load, which can be modelled as a mass supported by a spring, and is submerged in fluid providing appreciable damping. For comparison, the open-loop response to a near step in voltage is shown, which gives a similar change in position to that demanded for the closed-loop system. The closed-loop response is slightly faster but, most importantly, always exhibits negligible steady-state error.

Note 7.4
Closed-loop control requires integral actuation to avoid large steady-state error.

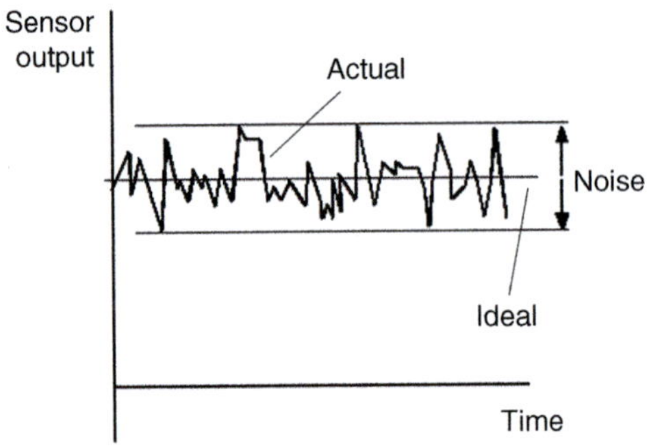

Figure 7.19 High-frequency noise.

7.8 Signal Conditioning for Piezoelectric Sensors

7.8.1 Op-Amp Filtering Circuits

An op-amp circuit is also used to interface with a piezoelectric device when it is used as a sensor rather than an actuator. The circuit is required to condition and scale the sensor output, providing a voltage in a suitable range (e.g. 0–5 V). An important part of conditioning is filtering out high-frequency noise (Figure 7.19), which might be due to electromagnetic interference, for example. Op-amp low-pass filtering circuits can be used to achieve this and are introduced in this section.

Figure 7.20 shows the simplest type of op-amp-based low-pass filtering circuit, a first-order filter. By summing the currents through C and R_2 and equating them to the current through R_1, we get the following differential equation:

$$C\frac{dV_o}{dt} + \frac{V_o}{R_2} = -\frac{V_i}{R_1} \tag{7.14}$$

Or, in terms of the differential operator 's', the transfer function relating input and output voltages is:

$$V_o = -\frac{R_2/R_1}{R_2Cs + 1}V_i \tag{7.15}$$

By putting $s = 0$, we find that the steady-state gain of the filter is $-R_2/R_1$. The time constant for the filter is R_2C seconds, and its cut-off frequency is the reciprocal of this, $1/(R_2C)$ rad/s. Figure 7.21 shows how, with rising frequency, the voltage is attenuated increasingly above the cut-off frequency, and the reduction in output magnitude is accompanied by increasing phase lag. This plot is normalised to give unity gain (0 dB) at low frequency.

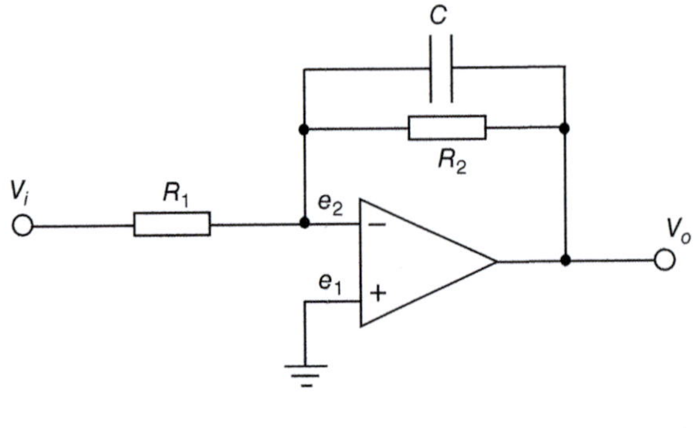

Figure 7.20 First-order low-pass filter (inverting).

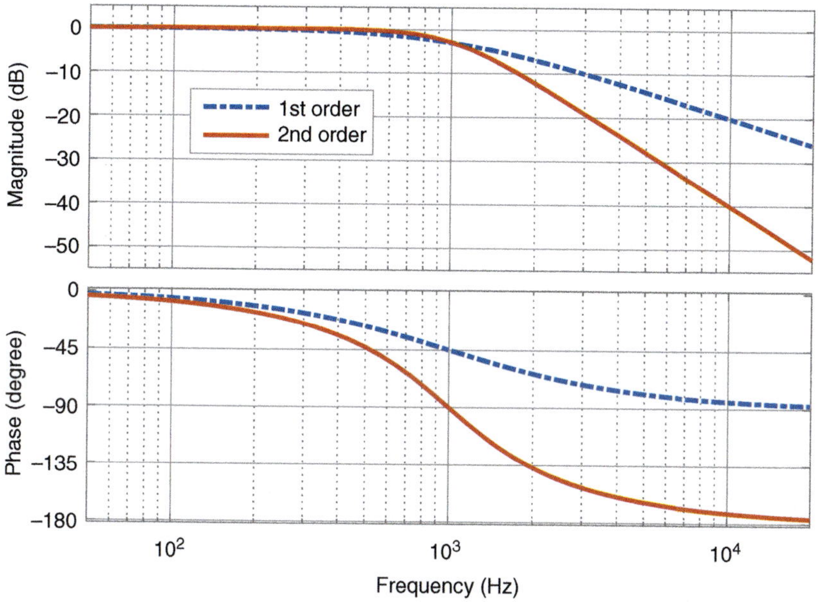

Figure 7.21 Bode plot of first- and second-order low-pass filters with a 1 kHz cut-off frequency.

Figure 7.22 Second-order low-pass filter (non-inverting).

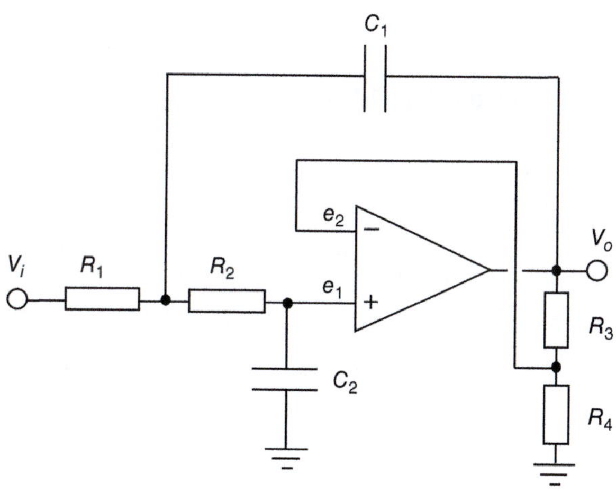

A second-order filter exhibits a more rapid attenuation above the cut-off frequency. This is shown in Figure 7.21 for a filter with a damping ratio of 0.7. The filter can be implemented as shown in Figure 7.22. The analysis to derive an input–output transfer function is along the same lines as for the first-order filter but is not included here.

An alternative type of filter that might be required is a high-pass filter. Rather than attenuating the high-frequency components of noisy signals, this is used for removing low-frequency errors, i.e. drift. Piezoelectric force or strain sensors suffer from drift because

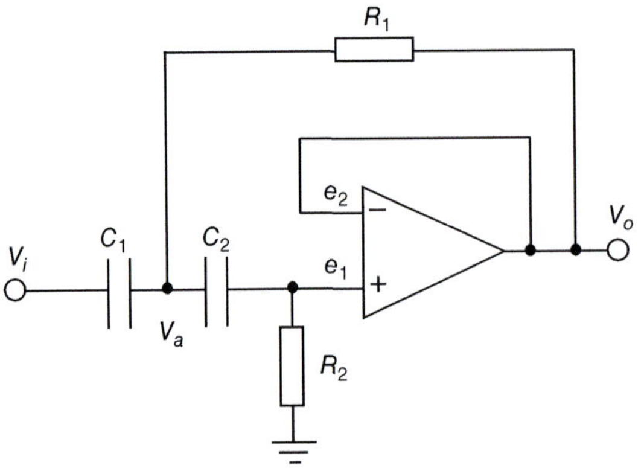

Figure 7.23 Second-order high-pass filter (non-inverting).

the electrical charge produced decays with time due to the internal leakage current. A second-order high-pass filter is shown in Figure 7.23. To analyse this circuit, notice that $e_1 = e_2 = V_o$, which determines the current through R_2. So:

$$\frac{V_o}{R_2} = C_2 \frac{d(V_a - V_o)}{dt} \tag{7.16}$$

The current through C_1 is the sum of currents through R_1 and R_2. So:

$$\frac{V_o}{R_2} + \frac{V_a - V_o}{R_1} = C_1 \frac{d(V_i - V_a)}{dt} \tag{7.17}$$

Differentiating Eq. (7.17):

$$\frac{1}{R_2}\frac{dV_o}{dt} + \frac{1}{R_1}\frac{d(V_a - V_o)}{dt} = C_1 \frac{d^2(V_i - V_a)}{dt^2} \tag{7.18}$$

By substituting for $\frac{d(V_a - V_o)}{dt}$ using Eq. (7.16) and then differentiating Eq. (7.16), multiplying through by C_1/C_2 and adding to Eq. (7.18), V_a can be cancelled out of Eq. (7.18), leading to:

$$\frac{d^2V_o}{dt^2} + \frac{1}{R_2}\left(\frac{1}{C_1} + \frac{1}{C_2}\right)\frac{dV_o}{dt} + \frac{1}{R_1R_2C_1C_2}V_o = \frac{d^2V_i}{dt^2} \tag{7.19}$$

or in transfer function form:

$$V_o = \frac{s^2}{s^2 + \frac{1}{R_2}\left(\frac{1}{C_1} + \frac{1}{C_2}\right)s + \frac{1}{R_1R_2C_1C_2}} V_i \tag{7.20}$$

Figure 7.24 shows the frequency response of an example second-order high-pass filter, with a cut-off frequency of 1 Hz and a damping ratio of 0.7. Low-frequency signal components are attenuated, and the steady-state (DC) gain is zero.

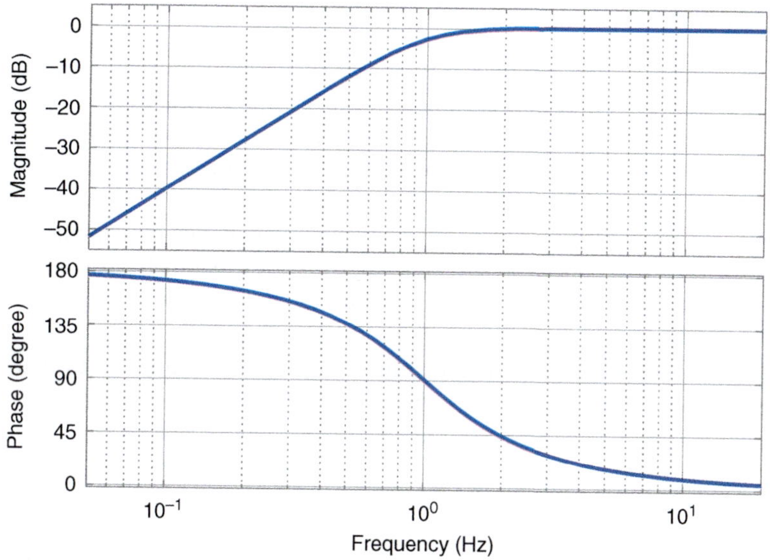

Figure 7.24 Second-order high-pass filter with a 1 Hz cut-off frequency.

Example Exercise 1

Design a first-order low-pass filter with a cut-off frequency of 13.3 Hz and a steady-state (DC) gain of −1. If a 1-μF capacitor is used, what resistor values are required?

Solution. Refer to Eq. (7.15). For a steady-state gain of −1, $R_1 = R_2 = R$, and so, $V_i = -\dfrac{1}{RCs+1}V_o$. The cut-off frequency should be $13.3 \times 2\pi = 83.6$ rad/s. The time constant is RC, and so, we want $RC = 1/83.6 = 0.012$ s. Hence, $R = 12\,\text{k}\Omega$.

Example Exercise 2

Piezoelectric accelerometers give an inaccurate measurement at low frequencies. Design an op-amp circuit which behaves as a second-order high-pass filter to attenuate the acceleration signal at low frequency; this should have a cut-off frequency of 1 Hz. All capacitors in the circuit should be the same size, and all resistors should also be the same size. What relationship between R and C values gives this cut-off frequency, and what is the damping ratio?

Answer. $RC = 1/(2\pi)$, $\zeta = 1$

Note 7.5

Analogue op-amp circuits can be used to reduce frequency-related errors from sensor signals, e.g. high-frequency noise or low-frequency drift. Digital filters can alternatively be used, but sampling of signals causes aliasing where high-frequency noise appears at lower frequencies, so analogue low-pass filtering cannot usually be avoided.

7.8.2 Signal Conditioning in Detail

A charge amplifier is presented in this section for interfacing to a piezoelectric sensor to give an output voltage in a suitable range for measurement. A model is presented for the amplifier, the sensor and the cable connecting to the two, so that the dynamic performance of the measurement system can be analysed. Note that piezoelectric sensors are not suitable for static (DC) measurements because the electrical charge produced decays with time due to the internal leakage current. However, they are well suited to dynamic measurement. A piezoelectric sensor can be modelled as a charge source with a parallel capacitance (C_p) and also a high resistance (R_p) in parallel to allow for the leakage current (see Figure 7.25). The circuit shown is a charge amplifier, using a TLV2771 op-amp, which has a particularly high input impedance. The non-inverting input is connected to half the supply voltage V_{cc}, so the inverting input is effectively tied to the same voltage. Due to this fixed voltage, the basic characteristic of the circuit, assuming R_f is large and R_i is small, is that any charge generated by the sensor goes to the charging capacitor C_f, and so, the output voltage is:

$$V_o \approx \frac{V_{cc}}{2} - \frac{1}{C_f} q_p \tag{7.21}$$

Resistor R_f bleeds the charge off capacitor C_f at a low rate to prevent the amplifier from drifting into saturation and ensures the output tends to the op-amp input voltage ($V_{cc}/2$) in the steady state. The resistor R_i provides electrostatic discharge protection.

A dynamic analysis of the circuit is given as follows. Both op-amp inputs are taken to be at a fixed voltage of $V_{cc}/2$. The current input to the amplifier is given by the rate of charge generation by the sensor, less the leakage current through R_p and any change in charge stored in the sensor and cable capacitance ($C_p + C_c$):

$$i_i = \frac{dq_p}{dt} - \frac{V_p - V_{cc}/2}{R_p} - \left(C_p + C_c\right) \frac{dV_p}{dt} \tag{7.22}$$

Considering the current through the amplifier input resistor R_i:

$$V_p = i_i R_i + V_{cc}/2 \tag{7.23}$$

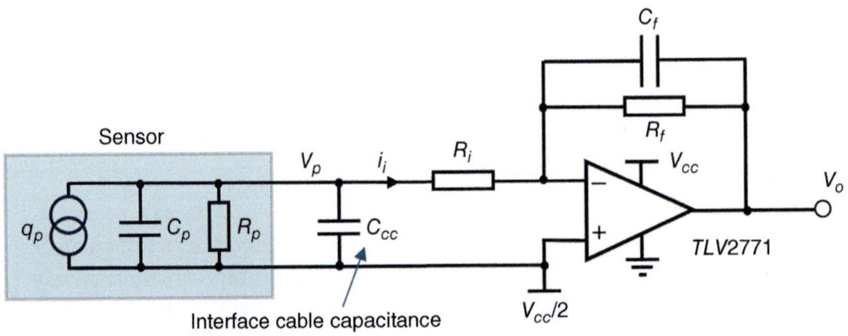

Figure 7.25 Charge amplifier for piezoelectric sensor signal conditioning (adapted from [8]).

So substituting for V_p in Eq. (7.22):

$$i_i = \frac{dq_p}{dt} - \frac{R_i}{R_p} i_i - R_i\left(C_p + C_c\right)\frac{di_i}{dt}$$

$$R_i\left(C_p + C_c\right)\frac{di_i}{dt} + \left(1 + \frac{R_i}{R_p}\right) i_i = \frac{dq_p}{dt} \tag{7.24}$$

or in terms of the differential operator s:

$$\left[R_i\left(C_p + C_c\right)s + \left(1 + \frac{R_i}{R_p}\right)\right] i_i = s q_p \tag{7.25}$$

The amplifier output voltage is determined by the current i_i, which is split between the feedback capacitor and resistor (C_f and R_f):

$$i_i = C_f \frac{d(V_{cc}/2 - V_o)}{dt} + \frac{V_{cc}/2 - V_o}{R_f}$$

$$R_f C_f \frac{dV_o}{dt} + V_o = \frac{V_{cc}}{2} - R_f i_i$$

$$\left(R_f C_f s + 1\right) V_o = \frac{V_{cc}}{2} - R_f i_i \tag{7.26}$$

Substituting for i_i in Eq. (7.26) using Eq. (7.25):

$$\left(R_f C_f s + 1\right) V_o = \frac{V_{cc}}{2} - R_f \frac{s}{R_i\left(C_p + C_c\right)s + \left(1 + R_i/R_p\right)} q_p$$

$$V_o = \frac{V_{cc}/2}{R_f C_f s + 1} - \left[\frac{1}{C_f\left(1 + R_i/R_p\right)}\right] \frac{s}{\left(s + \frac{1}{R_f C_f}\right)\left(\frac{R_i\left(C_p + C_c\right)}{1 + R_i/R_p} s + 1\right)} q_p \tag{7.27}$$

Equation (7.27) is the complete equation relating sensor charge to output voltage. However, other than when the amplifier is first switched on, V_{cc} is constant. In addition, the leakage path resistance R_p will be very much higher than R_i, so Eq. (7.27) can be simplified to:

$$V_o = \frac{V_{cc}}{2} - \frac{1}{C_f} \frac{1}{\left(1 + \frac{1}{R_f C_f s}\right)\left(R_i\left(C_p + C_c\right)s + 1\right)} q_p \tag{7.28}$$

The time constant represented by $R_f C_f$ needs to be large to extend the operational range of the amplifier in the low-frequency direction, and the time constant $R_i(C_p + C_c)$ needs to be small to extend the high-frequency range. Specifically, from Eq. (7.28), it can be shown that the amplitude of sinusoidal charge and voltage signals is related by:

$$|V_o| = \frac{1}{C_f} \frac{1}{\sqrt{\left(1 + \frac{1}{\left(R_f C_f \omega\right)^2}\right)\left(\left(R_i\left(C_p + C_c\right)\omega\right)^2 + 1\right)}} \left|q_p\right| \tag{7.29}$$

where ω is the sinusoidal frequency in rad/s. Thus, the range of satisfactory operation of the amplifier, in which Eq. (7.28) approximates Eq. (7.21), is given by:

$$\frac{1}{R_f C_f} << \omega << \frac{1}{R_i\left(C_p + C_c\right)} \tag{7.30}$$

or

$$\frac{1}{2\pi R_f C_f} << \text{frequency (Hz)} << \frac{1}{2\pi R_i\left(C_p + C_c\right)} \tag{7.31}$$

For example, consider a piezoelectric hydrophone with a combined sensor and cable capacitance of 4000 pF. With $R_i = 1\,\text{k}\Omega$, the high-frequency bandwidth is 40 kHz. With $R_f = 100\,\text{M}\Omega$ and $C_f = 100$ pF, the low-frequency cut-off is at 16 Hz. The gain of the amplifier is also determined by C_f; in this case, the output voltage is amplified by a factor of 40 compared to the equivalent open-circuit sensor voltage (i.e. $(C_p + C_c)/C_f$).

Note 7.6

Sensor charge amplifiers normally have a band-pass characteristic, i.e. very low and very high frequencies will be attenuated.

7.9 Concluding Remarks

Any piezoelectric device used as an actuator, sensor or energy harvester will require connection to a network of electrical and electronic components to manage the piezoelectric voltage and/or charge. Actuators are likely to be controlled by digital electronics, requiring amplification of control signal voltages to high-power signals, sometimes involving both high voltages (up to ~1000 V) and high current (several amps). Piezoelectric sensor power levels are lower, but careful conditioning of the signal to provide low-noise, low-drift measurements is equally important.

We have also seen that piezoelectric actuators alone do not give us the ability to accurately position a load. They need to be part of a closed-loop control system in which position is measured, and an appropriate closed-loop controller is deployed.

References

1 Robinson, S. (2006). *Driving Piezoelectric Actuators*. New York, USA: Power Electronics Technology April 2006.

2 Waanders, J.W. (1991). *Piezoelectric Ceramics: Properties and Applications*. Eindhoven: Philips Components.

3 Fleming, A.J. and Moheimani, S.O.R. (2004). Improved current and charge amplifiers for driving piezoelectric loads, and issues in signal processing design for synthesis of shunt damping circuits. *Journal of Intelligent Materials Systems and Structures* 15: 77–92.

4 Serridge, M. and Licht, T.R. (1987). *Piezoelectric Accelerometers and Vibration Preamplifiers: Theory and Application Handbook*. Denmark: K Larden & Son A/S.
5 Karpelson, M., Wei, G.-Y., and Wood, R. (2012). Driving high voltage piezoelectric actuators in microrobotic applications. *Sens. Actuators A: Phys.* 176: 78–89.
6 Lindner, D., Zhu, H., Song, C. et al. (2002). *Low Input Voltage Switching Amplifiers for Piezoelectric Actuators*. Industrial and Commercial Applications of Smart Structures Technologies: Proceedings of SPIE North American Symposium on Smart Structures and Materials.
7 Persson, J., Plummer, A., Bowen, C., and Brooks, I. (2015). Design and modelling of a novel servovalve actuated by a piezoelectric ring bender. In: *in ASME BATH 2015 Symposium On Fluid Power Motion Control*. Chicago: Oct 2015.
8 Karki, J. (2000). Signal conditioning piezoelectric sensors. Texas Instruments Application Report. SLOA033A – Sept 2000.

8

Case Studies

8.1 Introduction

Many applications have been proposed for piezoelectric sensing, actuation and energy harvesting. Some devices have been successfully commercialised, targeting specific applications, while others remain a matter of research and development. Examples and detailed studies are presented here, focusing on piezoelectric actuation and piezocomposite sensing.

8.2 Piezoelectric Valve Actuation

8.2.1 Internal Combustion Engine Fuel Injectors

Piezoelectric actuation of valves for controlling the flow of liquids and gases is an area which has been extensively investigated and commercially exploited for some applications. Often, the motivation for replacing electromagnetic valve actuation technology is to increase the speed of response.

In 2008, piezoelectric diesel fuel injectors became commercially available for internal combustion engines. Conventional injectors use an electromagnetic solenoid to exploit the power of the pressurised fuel to open the injector; this two-stage valve architecture is required, as a compact solenoid does not produce enough force to directly move the injector needle. The concept of a piezoceramic stack-actuated injector is shown in Figure 8.1; the stack would be energised to close the valve and de-energised to inject fuel. In the Delphi direct acting common rail system of 2008, the injector needle is moved by a stack via a hydraulic motion amplifier [1]. Figure 8.2 shows that there is an area ratio between the stack and the inner end of the injector needle, providing the amplification. Compared to solenoid injectors, the piezoelectric injector gives faster needle opening and closing, which is independent of the injection pressure, potentially leading to improved emissions and fuel consumption. The time for the complete injection cycle is reduced from 1.5 to 0.5 ms. Proportional control of needle lift is also possible to further improve engine performance. The physical construction of an example piezoelectric diesel fuel injector is shown in Figure 8.3.

Piezoelectric Materials: Properties, Applications and Devices, First Edition.
Christopher R. Bowen and Andrew R. Plummer.

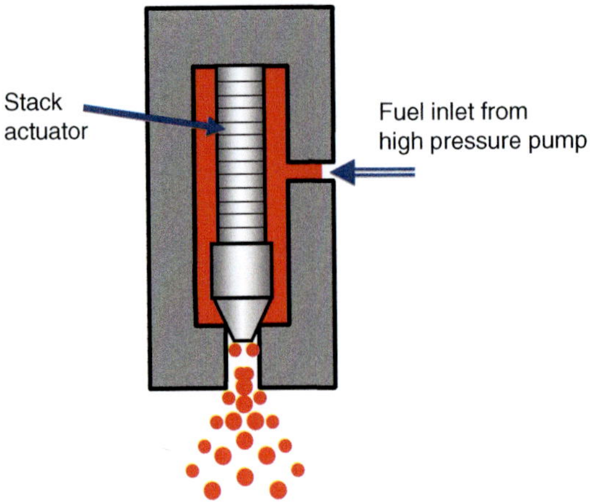

Figure 8.1 Piezoceramic fuel injector concept.

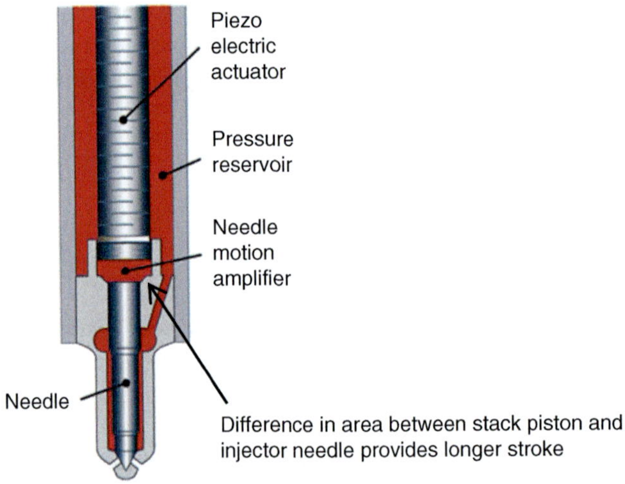

Figure 8.2 Delphi direct acting common rail fuel injector. *Source:* Schöppe et al. [1] / with permission of Springer Nature.

8.2.2 Hydraulic Servo Valves

The servo valve is the key component in electrohydraulic motion control systems, providing fast and accurate control of flow to hydraulic cylinders, and is used widely for actuation in aircraft, testing machines, manufacturing systems (hydraulic presses, injection moulding, etc.) and many other applications [3]. In a direct-drive, or single-stage, servo valve, a spool is positioned by an electrical actuator in proportion to the electrical input. The spool movement changes the size of the metering orifices, thus enabling the valve to control flow. In a two-stage servo valve, the spool is also positioned in proportion to the electrical input to the valve; however, the spool movement is achieved through internal hydraulic actuation. The most common servo valve design is the two-stage nozzle–flapper valve, where an electromagnetic torque motor moves a flapper that sits between a pair of nozzles, and the resultant

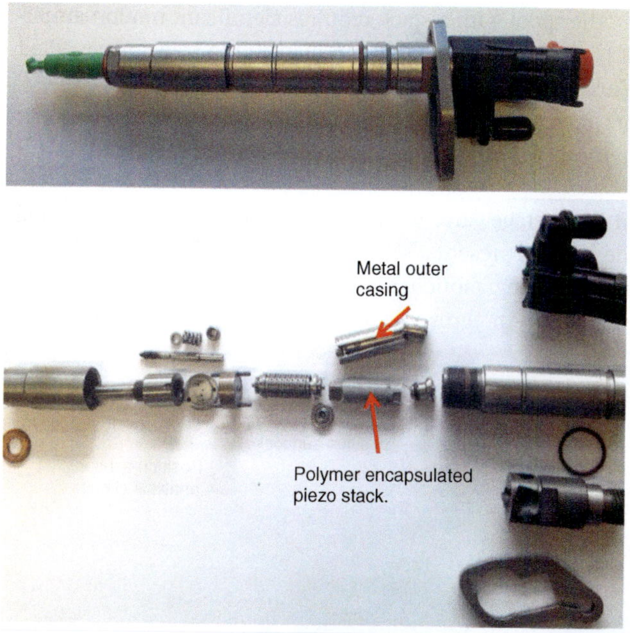

Figure 8.3 Bosch piezoelectric diesel fuel injector, complete (top) and disassembled (bottom). *Source:* [2] / Courtesy of Johan Person.

imbalance in the nozzle outflows creates a pressure drop across the ends of the spool. The consequent change in spool position is accurately regulated through either mechanical feedback to the torque motor via a feedback spring or electrical closed-loop control via a spool position transducer.

There have been many studies in replacing the torque motor or other electromagnetic actuators in a servo valve with a piezoelectric actuator. Due to small maximum strains, using a piezoelectric stack realistically requires motion amplification, even for the first-stage actuation (e.g. flapper movement of around 0.1 mm). Rectangular bending actuators can provide sufficient displacement but fairly small forces. Ring bender actuators provide sufficient displacement for the first-stage actuation and reasonable force levels (~10–100 N). Such benders are available with ceramic layers as thin as 20 μm, in which case, electrode voltages of around 50 V provide sufficient field strength. Issues that need to be considered in valve design are piezoelectric hysteresis, creep and the fact that the stack actuator length is temperature dependent. In addition, as the actuator behaves like a capacitor, the speed of response is generally constrained by the amplifier current limit.

In a 1955 survey of the then new field of servo valve technology [4], electromagnetic actuation is the focus for the electrical-to-mechanical conversion, but it is stated that 'piezoelectric crystals have been used on certain experimental models to obtain improved response. However, they have not been accepted to date because of high susceptibility to vibration, temperature changes and electrical noise and because of the difficulty in obtaining sufficiently large displacements from the crystals'. A patent for a piezoelectric valve was filed in 1955, covering a piezo-actuated flapper for a nozzle–flapper two-stage valve and also delivering fluid using an oscillating piezo-disc, i.e. a piezo pump [5].

In a single-stage valve, moving the spool with a stack requires significant motion amplification. In a valve described in [6], this is achieved with a hydrostatic transformer filled with silicone and a 40:1 piston area ratio. A bandwidth frequency of 270 Hz is achieved, and using two opposing actuators at either end of the spool reduces temperature sensitivity (Figure 8.4). An alternative method of amplification, using a lever, is reported in [7] (Figure 8.5).

Replacing the torque motor in a two-stage valve with a piezoelectric actuator is reported in a number of studies. In [8], the authors present a servo valve where a flextensional actuator (a stack in a flexing frame providing motion amplification) moves a flapper in a mechanical feedback valve (Figure 8.6). An aerospace servo valve, again with a feedback

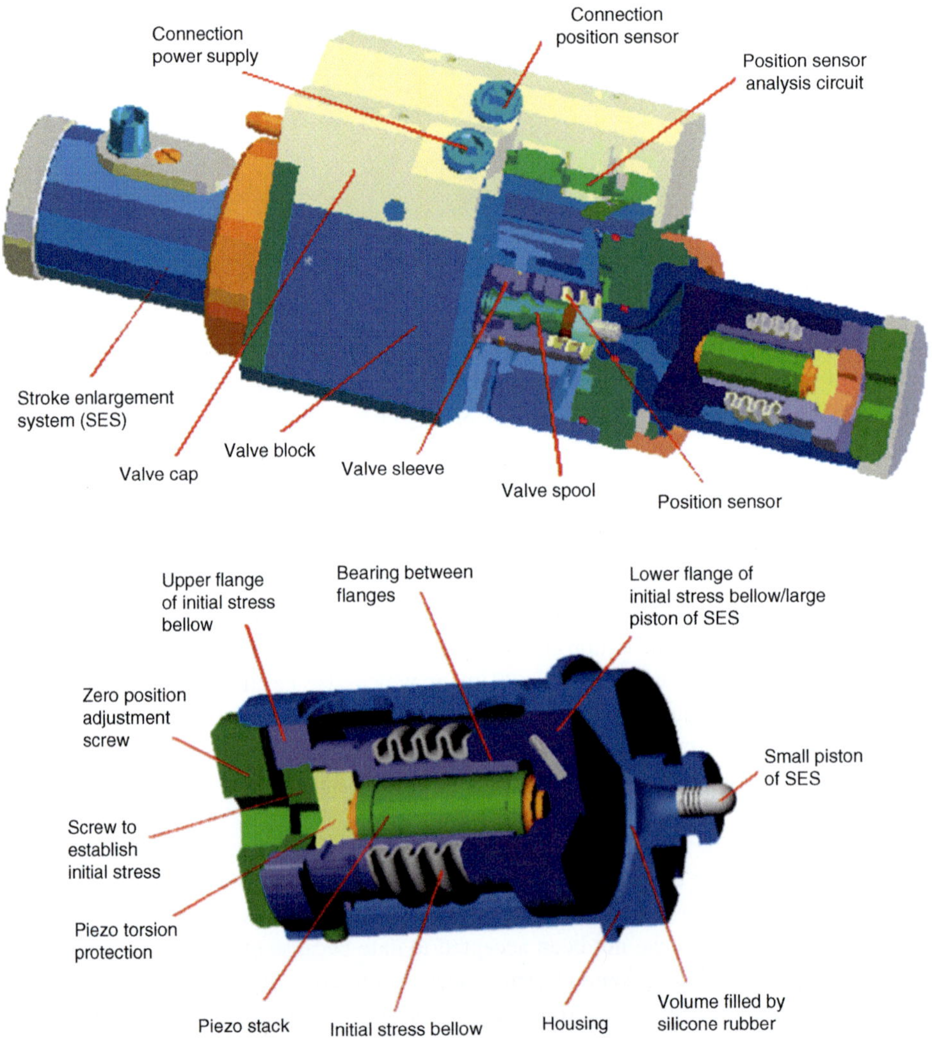

Figure 8.4 Valve (top) actuated by hydrostatically amplified piezoelectric stack modules (bottom). From [6].

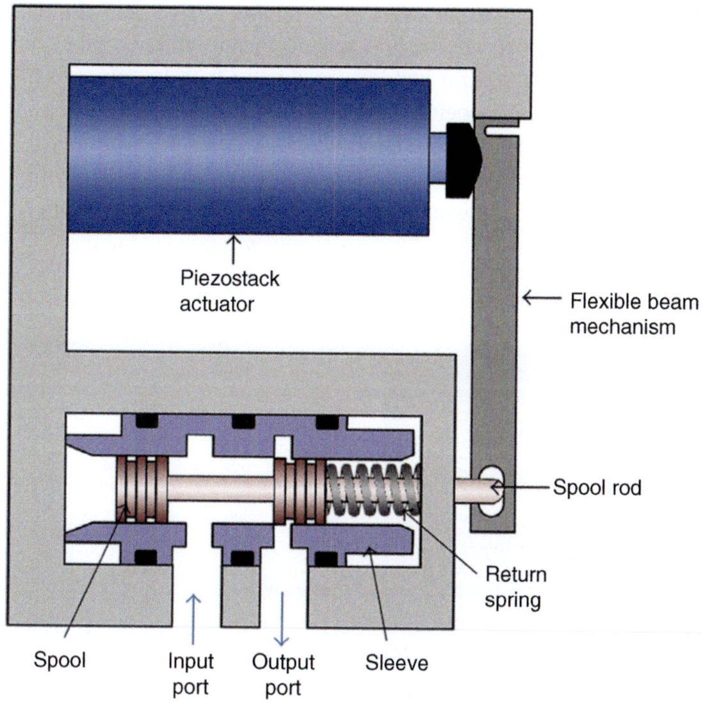

Figure 8.5 Spool actuation with mechanically amplified piezoelectric stack motion. *Source:* Jeon et al. [7] / with permission of Sage publications.

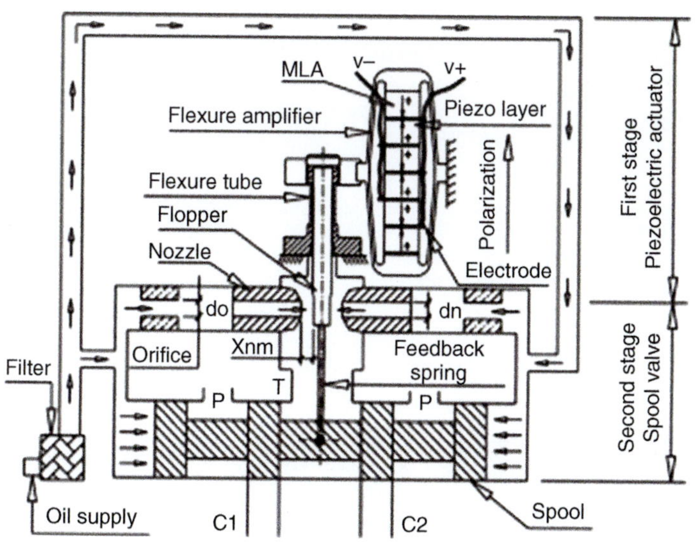

Figure 8.6 Piezo stack with flextensional amplification for a two-stage valve. *Source:* Karunanidhi and Singaperumal [8] / with permission of Sage publications.

wire, is presented in [9]. This uses a rectangular piezoelectric bender to move a deflector jet, arguing that the smaller flow forces experienced in a first-stage deflector jet (or jet pipe), as opposed to a nozzle–flapper, are more suited for bender use (Figure 8.7). In comparison with a torque motor, it is suggested that a piezoelectric bender may prove easier to manufacture and commission and give more repeatable performance. In a recent valve prototype, a ring bender is used as the first-stage actuator [10, 11]. This time, the first stage is a miniature spool, which gives the advantage of reduced leakage flow. Electrical spool position feedback is used (Figure 8.8).

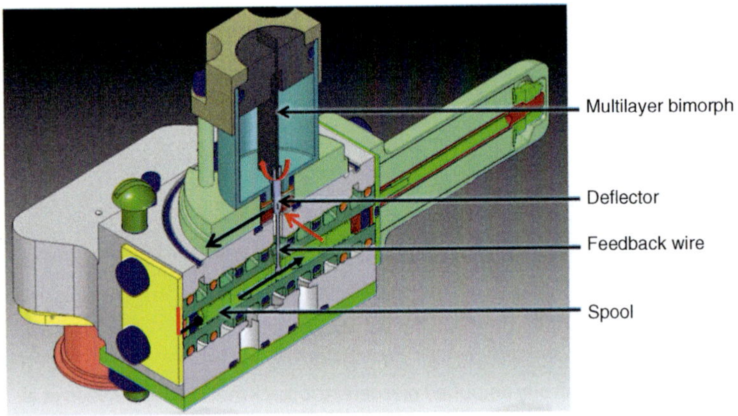

Figure 8.7 Piezoelectric rectangular bender deflector jet two-stage MFB valve. *Source:* Sangiah et al. [9] / with permission of Sage publications.

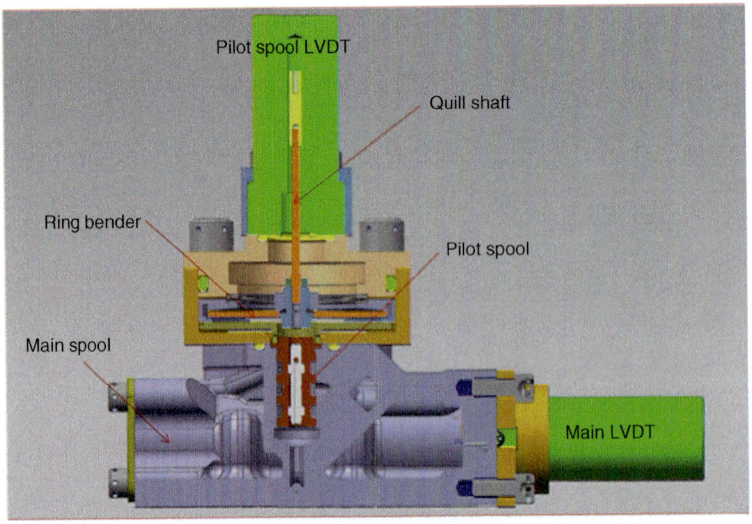

Figure 8.8 Piezoelectric ring bender-actuated pilot spool in two-stage EFB valve. *Source:* Persson et al. [11] / with permission of University of Bath.

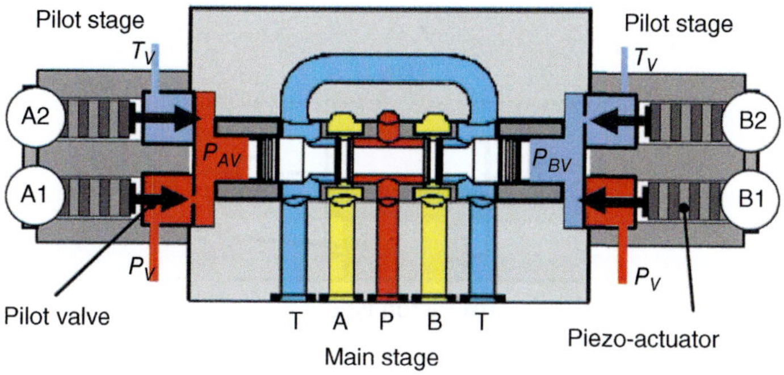

Figure 8.9 Independent piezo stack control for first-stage H-bridge orifices. *Source:* Reichert and Murrenhoff [12] / with permission of University of Bath.

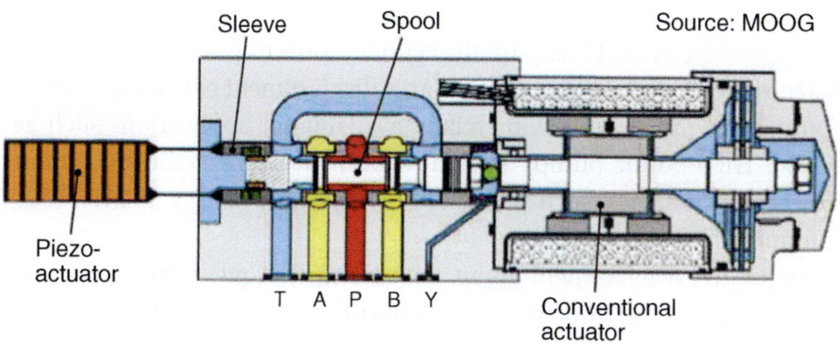

Figure 8.10 Dual-actuated valve, combining high-frequency (piezoelectric) and long-stroke (electromagnetic) actuators. *Source:* Reichert and Murrenhoff [12] / with permission of University of Bath.

Another piezo stack-actuated first-stage concept is described in [12]. As shown in Figure 8.9, the first stage consists of four controllable orifices forming an H-bridge; the orifices are modulated using stacks with a 40-μm stroke, and a bandwidth of over 1 kHz is achieved.

Figure 8.10 shows a novel concept for increasing the frequency response of a direct-drive valve. The spool-bushing sleeve is moved +/−20 μm using a stack, complementing the conventional +/−1-mm spool movement driven by a linear force motor. Thus, fine flow control can be achieved at much higher frequencies than the 60-Hz bandwidth of the conventional valve [12].

8.3 Piezoelectric Pumps

8.3.1 Introduction

A piezoelectric pump (commonly contracted to 'piezo pump') has a piezoelectrically actuated piston or diaphragm for drawing in and discharging fluid from a cylinder. A pair of non-return valves is used to create unidirectional flow. Figure 8.11 illustrates the concept, using either a diaphragm and bending actuator or a stack-driven piston.

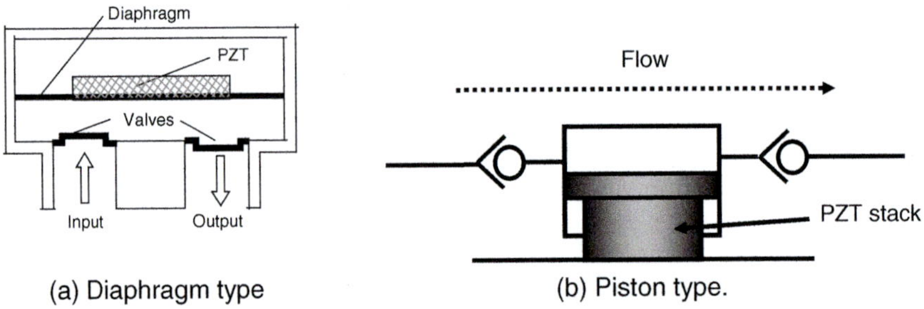

Figure 8.11 Piezoelectric pump; commonly contracted to 'piezo pump'.

The non-return valves perform hydraulic rectification, where fluid is forced out of the pumping chamber through an open outlet at high pressure. On the return stroke, the outlet shuts off, and the inlet opens, drawing fluid into the chamber. Continued cycling creates a flow. For actuation, the flow can be fed to a hydraulic cylinder; thus, repeated small displacements of the piezoelectric stack are converted into a large displacement of the load. Small, low-pressure piezo pumps (<1 mL/s) are commercially available for applications such as medical dispensing [13]. High-power pumps have also been developed; the maximum reported pressure differential that has been generated by a piezo pump is 10 MPa (100 bar), achieved at a pumping frequency of 2 kHz [14].

Neglecting losses, in a positive displacement pump, the volume flow rate increases in proportion to the pumping frequency; therefore, researchers have experimented with increasing the driving frequency into the kHz range. However, the flow rate has usually been less than expected. Chaudhuri *et al.* attribute the decrease in output of a piezo pump above 500 Hz to the inertia of the fluid [15, 16]. It is found in [17] that vortex formation at the outlet, or just inside the inlet, will cause pressure losses that will become greater with a higher flow rate. The energy efficiency of a piezo pump in [18] was found to be 7% using an electrostrictive lead magnesium niobate (PMN) stack or 1% using a piezoelectric lead zirconate titanate (PZT) stack. An experimental efficiency of 7% was also obtained by a separate group [19].

The inlet and outlet of the chamber can be controlled by means of passive non-return valves, as shown in Figure 8.11, or actively controlled valves, perhaps piezoelectrically actuated [10–22]. The additional complexity of using active valves might be justified in terms of faster opening – not relying on the valve's passive dynamics – and also being able to change the direction of flow by swapping the phasing of the two valves. However, active valves inevitably consume power, and this becomes prohibitive at high frequency [14].

8.3.2 Piezo Pump Example

A design of piezo pump with passive non-return valves (also known as check-valves) is shown in Figure 8.12 and pictured in Figure 8.13 [23]. The pump consists of two inlet check-valves, one outlet check-valve, a piezoelectric ring stack from Noliac with a nominal free displacement of 65 μm and a blocking force of 2700 N, and a 20-mm-diameter piston and housing. The piezoelectric stack is pre-tensioned by tightening a nut against a spring washer, which stresses the piston rod and compresses the stack. Once pre-tensioned, the

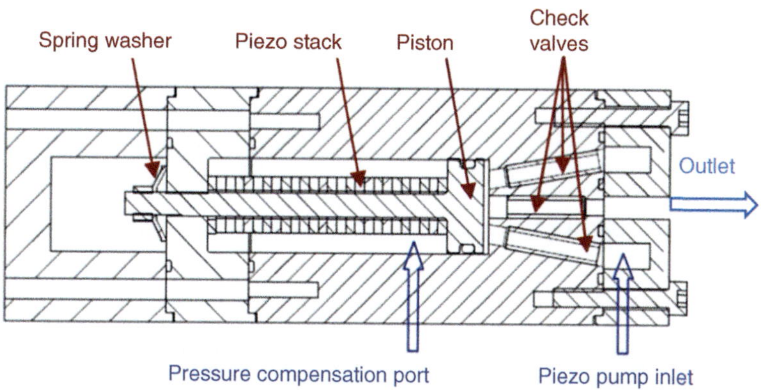

Figure 8.12 Piezo pump design. *Source:* Henderson et al. [23] / with permission of ASME.

Figure 8.13 Photograph of the pump; pumped flow displaces a hydraulic actuator on the right. *Source:* Henderson et al. [23] / with permission of American Society of Mechanical Engineers (ASME).

maximum working piezo-displacement is around 40 μm. The pump is designed to be very stiff structurally, and also hydraulically by having a small trapped volume, in order to minimise compressibility losses.

The pump flow is directed into a hydraulic actuator with a 15 mm piston diameter (area 1.77 cm^2) so that the flow can be deduced from the actuator motion. The actuator has a return spring requiring a force of 25.0 N to overcome; thus, the pump needs to develop a pressure of at least 1.4 MPa (14 bar) to move the actuator piston. Figure 8.14 shows the actuator motion for a sinusoidal piezo-driving voltage with a range of frequencies; it also shows the pump piston displacement, which has a stroke of about 30 μm when under load.

The flow and hence velocity are frequency dependent, but an increase in the piezo pump frequency above 90 Hz does not bring any velocity increase. The velocities at 90 Hz and 150 Hz are the same, and at 120 Hz, it is even slightly slower than at 90 Hz. The performance degradation above 90 Hz is caused by the limited check-valve speed of

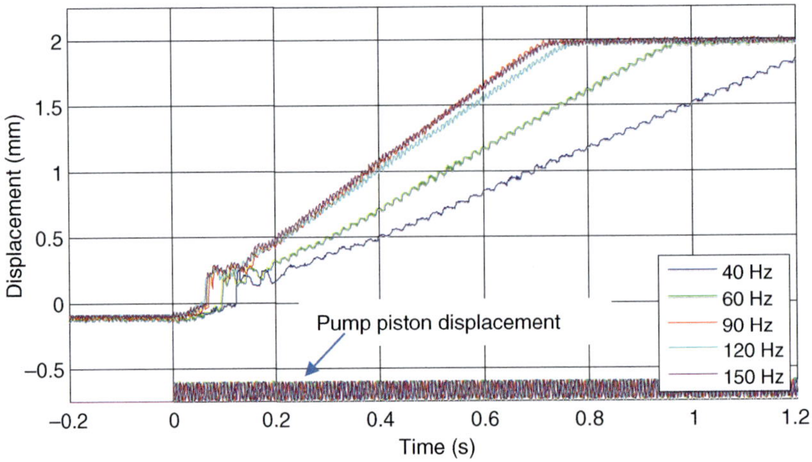

Figure 8.14 Measured actuator movement at different pump frequencies. *Source:* Henderson et al. [23] / with permission of ASME.

response. This performance drop-off can be quantified in terms of volumetric efficiency as shown below.

The theoretical average actuator velocity, u_t, depends on the stroke and the frequency of the piezo pump:

$$u_t = \frac{Q}{A_{act}} = \frac{A_{pump}}{A_{act}} \hat{X}_{pump} f \tag{8.1}$$

where Q is the average pump volume flow-rate, A_{act} and A_{pump} are the actuator and pump piston areas, respectively, $\hat{X}_{pump}$ is the stroke of the piezo pump, and f is the driving frequency. With diameters of 20 mm and 15 mm, respectively, the pump/actuator piston area ratio is 1.7778. The volumetric efficiency of the pump is given by

$$\eta_{vol} = \frac{u_m}{u_t} \tag{8.2}$$

where u_m is the experimentally measured average actuator velocity. Table 8.1 lists the ideal and actual actuator velocities for different pump frequencies, assuming a pump stroke of 30 μm, together with the calculated volumetric efficiencies. This clearly shows a dramatic drop in efficiency with frequency, which is mostly due to the check-valve response time.

Table 8.1 Theoretical and measured actuator velocities for different pump frequencies.

f (Hz)	u_m (mm/s)	u_t (mm/s)	η_{vol} (%)
40	1.667	2.133	78.1
60	2.227	3.2	69.6
90	2.959	4.8	61.6
120	2.725	6.4	42.6
150	2.882	8	36.0

8.4 Vibration Control of Flexible Structures

8.4.1 Smart Structure

Structures in dynamic mechanical systems are becoming lighter to reduce weight and inertia, improve compactness and reduce cost. However, this is accompanied by a reduction in the stiffness of these structures and, consequently, a reduction in the frequencies at which structural resonances occur. To counteract this trend, an ability to actively control vibration is attractive.

The smart structure approach to vibration control consists of embedding actuators and sensors within the structure. A microprocessor provides the intelligence to process sensor signals and drive the actuators, such that the structural dynamics are altered in a desired way. To be of practical use for complex structures, the controller should be simple to tune or have the ability to tune itself and not be dependent on the provision of a complex dynamic model of the structure derived from physical analysis.

In this case study, the vibration control of a simple cantilevered beam is studied, as shown in Figure 8.15 [24]. When perturbed, the beam exhibits significant vibration in the horizontal plane. Two strips of PZT piezoceramic are bonded near the clamped end of the beam and electrically connected in parallel. A full-bridge strain gauge arrangement is attached to the beam close to the piezoelectric actuators. An analogue interface card is used to monitor the amplified strain signal and transmit the control signal to a voltage amplifier, which drives the piezoelectric actuators in the range of ±90 V. Specific details of the experimental equipment are listed in Table 8.2.

Figure 8.16 shows the frequency response of the beam at the strain gauge sensor and at the tip, where a laser displacement sensor is used to monitor the displacement. Note that the strain gauge sensor happens to be near a point of zero bending for the second mode shape (45 Hz). Figure 8.17 (left) shows the free response of the beam following release from a strained position.

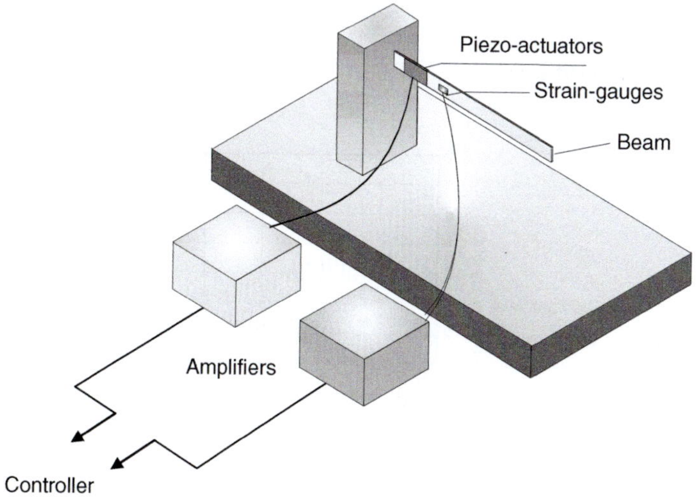

Figure 8.15 Experimental smart structure.

Table 8.2 Specification of the smart structure.

Beam	Dimensions: Material:	350 × 24 × 1 mm Aluminium
Piezoelectric actuator	Dimensions: Material: Piezoelectric strain coefficient (d_{31}): Young's modulus: Maximum safe voltage: Amplifier gain:	72 × 24 × 0.5 mm PZT, Navy Type II -190×10^{-12} m/V 4.9×10^{10} N/m^2 90 V 20
Sensor (strain gauges in full bridge)	Gauge factor and resistance: Location on the beam: Amplification:	2.1, 120 Ω 100 mm from the clamp 1200
Real-time controller	Sample rate:	100 Hz

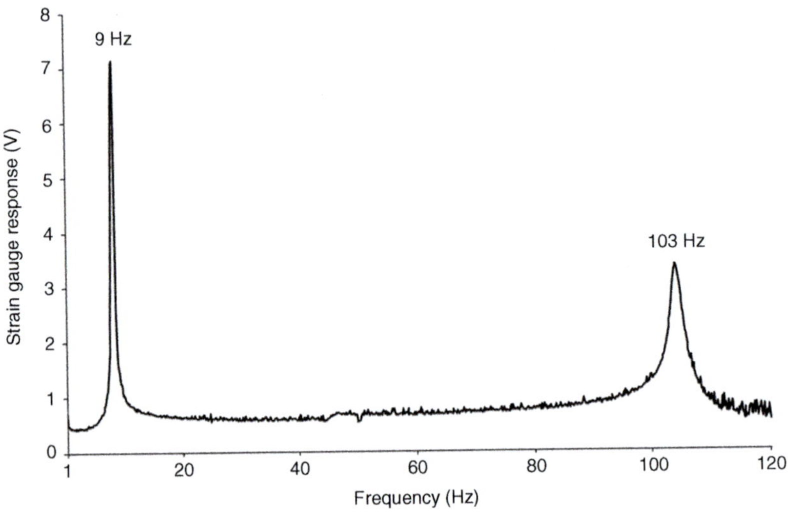

Figure 8.16 Frequency response of smart structure.

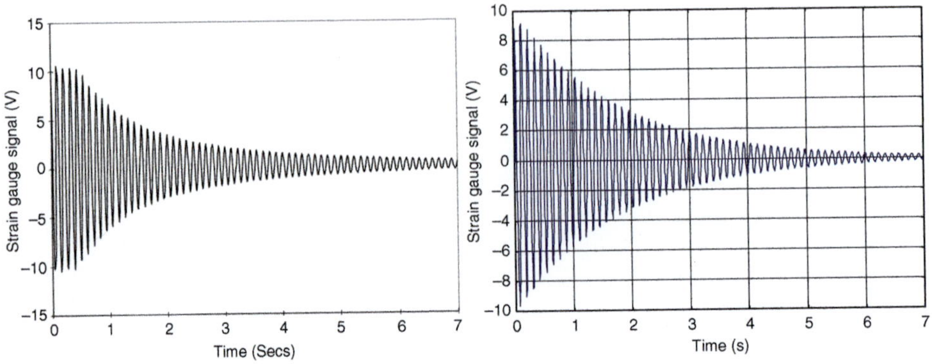

Figure 8.17 Free response of smart structure, measured (left) and simulated (right).

8.4.2 Dynamic Modelling

To help understand the behaviour of the smart structure, a model of the system is required. If y is a displacement-related measurement on a structure (in this case, strain), and f is a force applied by the piezoelectric actuator, these variables can be related by a sum of second-order transfer functions, each one representing one mode of vibration:

$$y = \left(\frac{\beta_1}{s^2 + \alpha_{11}s + \alpha_{10}} + \frac{\beta_2}{s^2 + \alpha_{21}s + \alpha_{20}} + \cdots + \frac{\beta_n}{s^2 + \alpha_{n1}s + \alpha_{n0}}\right)f \tag{8.3}$$

where s is the differential operator, and α_{ij} and β_i are the constants. In this case, the first mode of vibration is at a substantially lower frequency than the other modes, and so the model can be approximated by replacing the higher modes with their steady-state gains:

$$y = \left(\frac{\beta_1}{s^2 + \alpha_{11}s + \alpha_{10}} + C\right)f \tag{8.4}$$

where

$$C = \sum_{i=2}^{n} \frac{\beta_i}{\alpha_{i0}} \tag{8.5}$$

As the strain gauge is not co-located with the actuator, a constant force gives zero strain, so the steady-state gain is zero. Thus:

$$\frac{\beta_1}{\alpha_{10}} + C = 0 \tag{8.6}$$

Combining (8.4) and (8.6) gives:

$$y = \frac{Cs(s + \alpha_{11})}{s^2 + \alpha_{11}s + \alpha_{10}}f \tag{8.7}$$

In the frequency range of interest, the relationship between actuator control signal (u) and output force (f) can be approximated by a gain and a short delay:

$$f = Pe^{-ds}u \tag{8.8}$$

So (8.7) can be written in terms of u, giving the plant transfer function:

$$y = \frac{C_1 s(s + \alpha_{11})e^{-ds}}{s^2 + \alpha_{11}s + \alpha_{10}}u \tag{8.9}$$

By inspecting the free and forced responses of the system, it is possible to estimate the parameters in this transfer function as $\alpha_{10} = 2980$, $\alpha_{11} = 1.125$, $d = 0.01$ s and $C_1 = 0.12$. The first two parameters equate to a natural frequency of 54.6 rad/s and a damping ratio of 0.010. Figure 8.17 (right) shows that the free response of the model is very similar to that of the real system.

8.4.3 Derivative Feedback Control

The derivative of the strain gauge signal is an indication of the velocity close to the piezo-ceramic actuators, and, to a first approximation, the force output of the actuators is

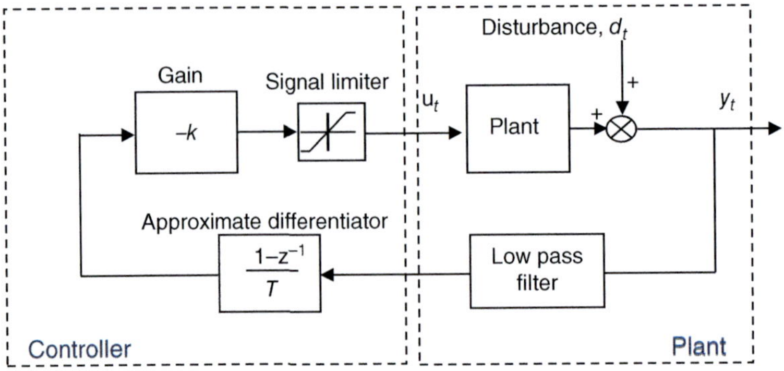

Figure 8.18 Derivative feedback controller (sample interval T).

proportional to the drive voltage. Thus, assuming the objective of the control system is to increase damping, using negative derivative feedback is a reasonable choice. Such a control scheme is shown in Figure 8.18. A 40-Hz low-pass filter is employed to attenuate noise. The saturation block is implemented in software and is set to ±4.5 V, limiting the output of the actuator voltage amplifier to ±90 V.

With a feedback gain of $k = 0.02$, the settling time is significantly reduced compared to the free response, as shown in Figure 8.19a. Increasing the feedback gain causes limit cycling. Figure 8.19b shows the response with $k = 0.06$. Figure 8.20 shows that simulating the system with the plant model of Eq. (8.7) predicts similar responses. The limit cycling is due to the saturation of the control signal voltage.

In conclusion, piezoelectric actuators can be used for closed-loop vibration control of flexible structures. As shown in the example, even a simple closed-loop controller can be effective, but care must be taken in its design. Typically, a gain which is too high causes instability. Stability analysis shows that the control signal limitation (saturation) plays a key role in determining the stable range of gains [24].

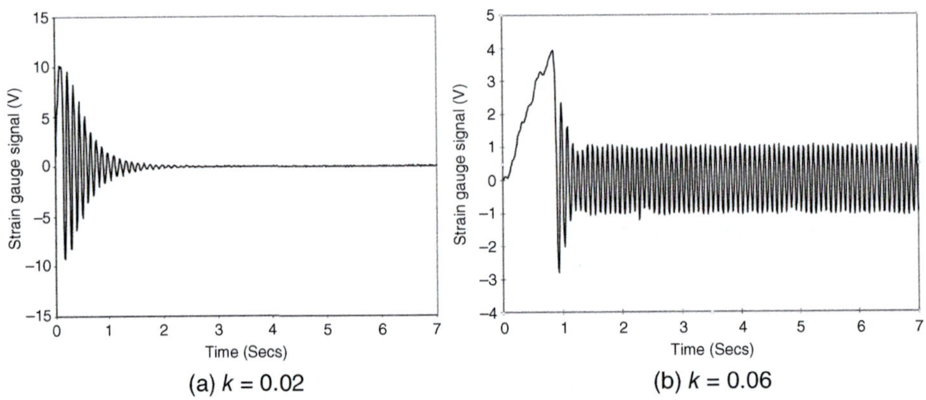

Figure 8.19 Response with a derivative feedback controller.

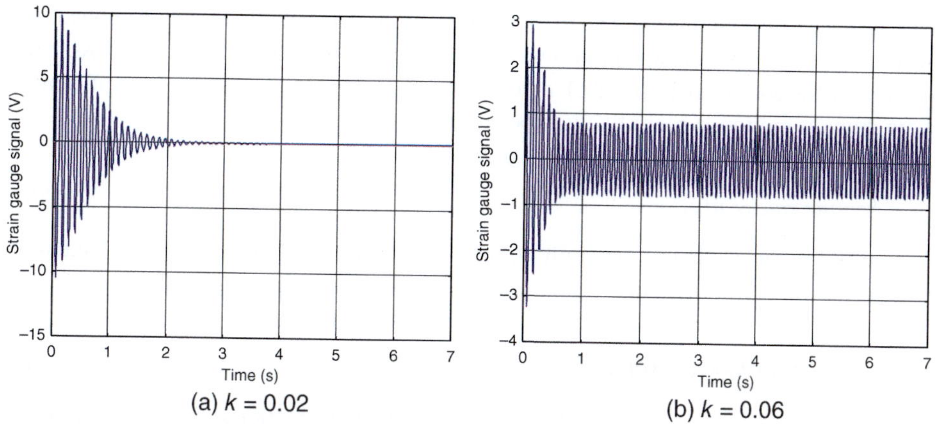

Figure 8.20 Simulated response with a derivative feedback controller.

8.5 A Case Study of Actuator Self-Heating

The ferroelectric hysteresis exhibited by a piezoelectric actuator is associated with an energy loss per electric field cycle that can lead to a degree of self-heating. Thus, rapid cycling of an actuator when operating at high frequencies can result in a dangerously high temperature rise, and it is important to be able to predict this heating characteristic. In this example, we will consider a multilayer actuator (MLA) based on thin piezoelectric layers (typically 0.1 mm or even lower). Due to the small distance between the electrodes, large electric fields and large displacements are generated at low drive potentials. The ability of an MLAs to operate at low voltages enables them to be used in a wide range of applications; this includes precision positioning, diesel fuel injection, printer heads, vibration control and optical instruments [25–27]. In an effort to optimise the displacement, 'soft' ferroelectric compositions are typically employed, such as those outlined in Chapter 3 [28, 29]. These soft ferroelectric materials exhibit relatively high d_{ij} piezoelectric strain coefficients, a measure of the strain per unit electric field. However, due to the domain motion and ferroelectric hysteresis, as outlined in the discussion of the polarisation–electric field loops and strain–electric field loops in Chapter 3, they also exhibit some form of hysteresis per electrical cycle. The existence of these losses leads to a degree of self-heating; this can be of particular concern if the actuator is being driven by a cyclic field. The temperature rise will depend on the specific operating conditions, where parameters such as the electric field amplitude, drive frequency, the properties of the material and the local heat transfer conditions are particularly relevant [30]. The mechanisms by which ferroelectric hysteresis losses are produced have been outlined in the literature, where domain wall motion and dielectric loss/conduction are important [31].

8.5.1 Model for Temperature Increase Due to Hysteresis

We will outline a simple model to predict self-heating [32]. If a piezoelectric actuator is subject to relatively large electric fields (e.g. in excess of 1 kV/mm and approaching the coercive field), ferroelectric domains are able to move with the electric field that is being applied. As a result, hysteresis losses from domain wall motion can be the primary source of loss, particularly for the soft ferroelectric PZT materials with high piezoelectric strain coefficients, where domain motion is more likely to occur [31–33].

The internal rate of actuator heating is given by

$$\textit{heating rate} \propto ufv_e \tag{8.10}$$

where u is the dielectric loss per unit volume per driving cycle (J/m^3) for the actuator material, f is the operating frequency (Hz), and v_e is the active volume (m^3). The rate of heat loss from the piezoelectric device to its surroundings is

$$\textit{rate of heat loss} \propto kA(T - T_0) \tag{8.11}$$

where k is defined as the overall heat transfer coefficient (W/m^2/K), A is the total surface area of the actuator (m^2), T_0 is the initial temperature (K), and T is the temperature at time t (K). The difference between the rates of heat generated and lost represents the rate of increase in retained heat energy; therefore:

$$ufv_e - kA(T - T_0) = v\rho c\frac{dT}{dt} \tag{8.12}$$

where v is the total volume, ρ is the density, and c is the specific heat capacity (J/kg/K) of the material [33].

A final, steady-state temperature rise ΔT is achieved when the rate of change in temperature is zero, namely when the rate of heat that is generated and the rate of heat lost are equal; therefore:

$$\Delta T = \frac{ufv_e}{kA} \tag{8.13}$$

Equation (8.13) indicates that any increase in temperature of a piezoelectric actuator, due to self-heating, will increase as the frequency (f) of operation increases; likewise, the temperature increase will increase with the level of ferroelectric hysteresis loss per cycle (which is associated with the 'softness' of the ferroelectric materials; see Chapter 3) and the amount of piezoelectric material that is being employed (v_e). As the driving field is increased, the degree of domain motion in the piezoelectric material will increase, and this will ultimately increase the amount of hysteresis losses per cycle (u) and therefore the temperature rise of the actuator; see Eq. (8.13). Equation (8.13) also indicates that a large total actuator area and a high heat transfer coefficient (namely good cooling conditions) will reduce the temperature increase, for example, by the application of forced airflow.

We can solve (8.12) to determine the actuator temperature as a function of time:

$$T = T_0 + \Delta T\left(1 - e^{-\frac{t}{\tau}}\right) \tag{8.14}$$

where the time constant is given by

$$\tau = \frac{v\rho c}{kA} \tag{8.15}$$

8.5.2 Testing Actuator Self-Heating

The influence of the driving field and operating frequency on the heating of an MLA is now studied. Overheating of the piezoelectric actuator can result in degradation or loss of actuator performance, for example, due to depolarisation or diffusion of electrode ions and oxygen vacancies due to the temperature rise [34–36]. Depolarisation due to heating will also be more likely for ferroelectric materials with a lower Curie point (T_c). The operational parameters for the actuator under test are given in Table 8.3, and the density and specific heat capacity of the PZT ferroelectric material have been used here. The presence of an electrode material (Ag–Pd alloy) is neglected since its volume is relatively low (< 10%).

Figure 8.21 is an image of the test configuration, and the MLAs are sandwiched between electrically insulating ceramic plates that are on a large steel base and cover (which act as large thermal masses); the upper plate and cover are not shown in Figure 8.21.

Table 8.3 Piezoelectric actuator dimensions and material properties.

Actuator dimensions	$10\,\text{mm} \times 10\,\text{mm} \times 3\,\text{mm}$
Actuator total surface area (A)	$320\,\text{mm}^2$
Number of piezoelectric layers	29
Thickness of piezoelectric layers	$90\,\mu\text{m}$
Active (piezoelectric) volume (v_e)	$8 \times 8 \times 2.8\,\text{mm}^3$
Density of lead zirconate titanate (ρ)	$7500\,\text{kg/m}^3$ [37]
Specific heat capacity of lead zirconate titanate (c)	420 J/kg/K [38]

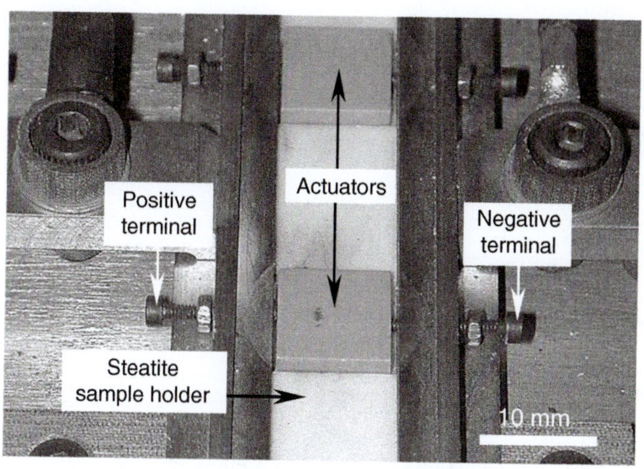

Figure 8.21 Position of piezo-actuators in the experimental setup, without cover-plate. *Source:* Pritchard et al. [32] / with permission of Elsevier.

The MLAs were electrically actuated by sinusoidal AC voltages at various frequencies and amplitudes, with a direct current bias applied to maintain a unipolar drive; this was selected to prevent depolarisation of the actuator. Figure 8.22a is the temperature–time response of the actuators when actuated at 50 Hz for applied voltages of 0–100 V, 0–150 V, 0–200 V and 0–250 V that correspond to applied electric fields of 1.1–2.8 kV/mm.

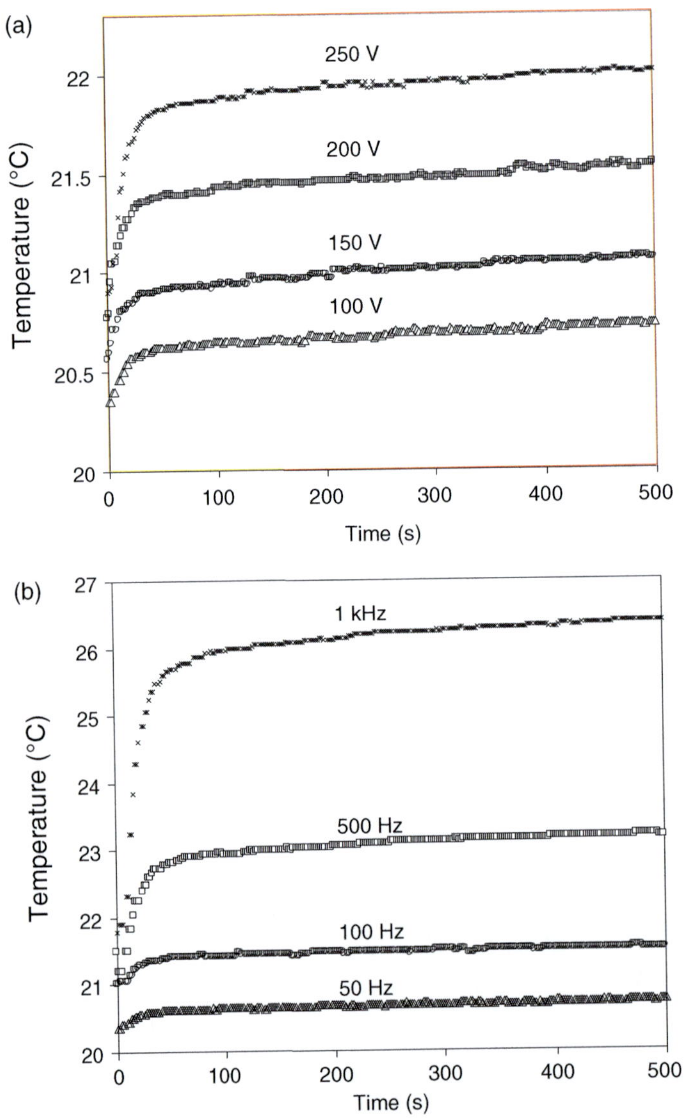

Figure 8.22 Temperature response of actuator (a) actuated at 50 Hz for a range of maximum voltages and (b) 100 V (1.1 kV/mm) peak at various frequencies. *Source:* Pritchard et al. [32] / with permission of Elsevier.

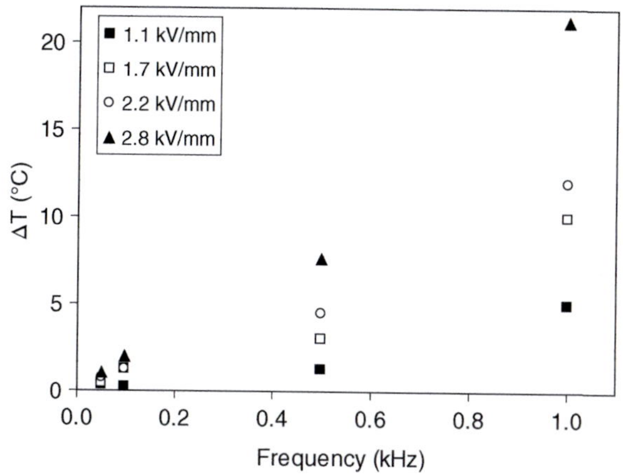

Figure 8.23 Temperature increase as a function of drive frequency for a range of electric fields. *Source:* Pritchard et al. [32] / with permission of Elsevier.

When examining the temperature–time response, initially at time $t = 0$, a potential is applied to the piezoelectric actuator, and there is an increase in temperature with time, which eventually reaches a constant value at longer periods, typically over 100 s. Figure 8.22b provides the temperature profile for a fixed electric field at a range of operating frequencies. An overall summary of the measured temperature rise for the actuators at the range of frequencies and driving fields tested is outlined in Figure 8.23. It can be observed that the temperature increase due to self-heating (ΔT) is approximately proportional to frequency; this is in good agreement with Eq. (8.13). The temperature rise also increases with an increase in the magnitude of the electric field; this is due to the larger hysteresis loss per cycle (u) as a result of a greater degree of domain motion [31, 32, 39].

8.5.3 Comparing Actuator Test Data with Expected Behaviour from Model

A typical example of temperature response of the piezoelectric actuator is provided in Figure 8.24 when the frequency is 1 kHz and the maximum applied voltage is 150 V. Equation (8.14) is then fitted to the experimental data points, which enables the time constant (τ) to be determined. The values of τ for the different frequencies and voltages allow the heat transfer coefficient, k, to be determined using Eq. (8.15); this is achieved using the v, ρ, c and A of the actuator provided in Table 8.3, as shown in Figure 8.25.

When piezoelectric actuators were simply hanging in the air with limited opportunity for cooling, the heat transfer coefficient was relatively small, between 20 and 40 $W/m^2/K$ [33]. However, in this case, the two large faces of the actuator are both in contact with 10-mm-thick ceramic plates that have a thermal conductivity of ~2.3 W/m/K. If we assume that the steel encasing the steatite ceramic forms a large heat-sink of constant temperature, this leads to a heat transfer coefficient for these faces of 230 $W/m^2/K$. If the edges of the actuator, total area of 120 mm^2, are considered to be in free air at 30 $W/m^2/K$, the area-weighted average heat transfer coefficient is 155 $W/m^2/K$; such a value is consistent with experimentally measured values.

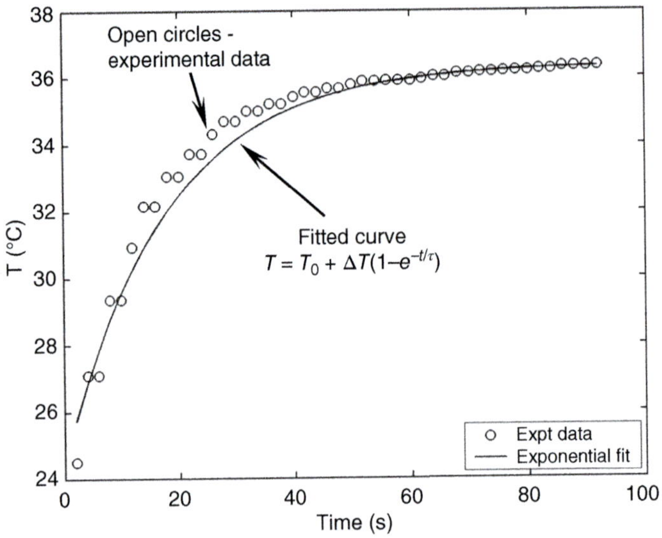

Figure 8.24 Model and measured temperature–time response of actuators at an AC frequency of 1 kHz and a maximum voltage of 150 V. *Source:* Pritchard et al. [32] / with permission of Elsevier.

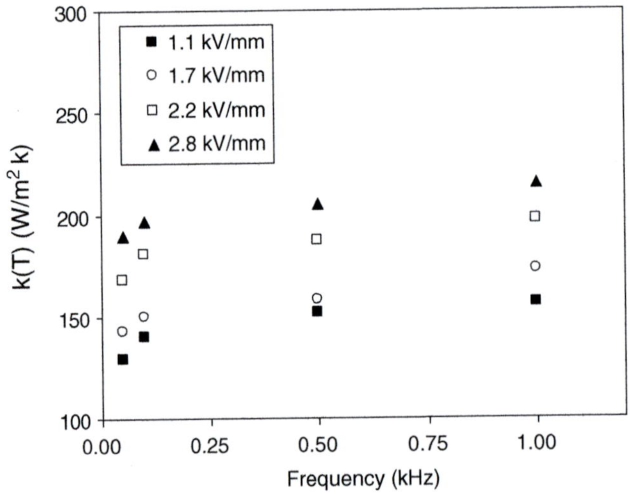

Figure 8.25 Estimated heat transfer coefficient values, based on fitting data to Figure 8.4. *Source:* Pritchard et al. [32] / with permission of Elsevier.

When the heat transfer coefficient is estimated from the time constant, the hysteresis loss per cycle and per unit volume (u) is calculated from the steady-state temperature rise via Eq. (8.13). Figure 8.26 outlines the change in hysteresis loss, u, with the magnitude of the electric field for a range of AC frequencies. This total loss is observed to increase with the electric field, but there is no clear correlation with frequency. The highest field

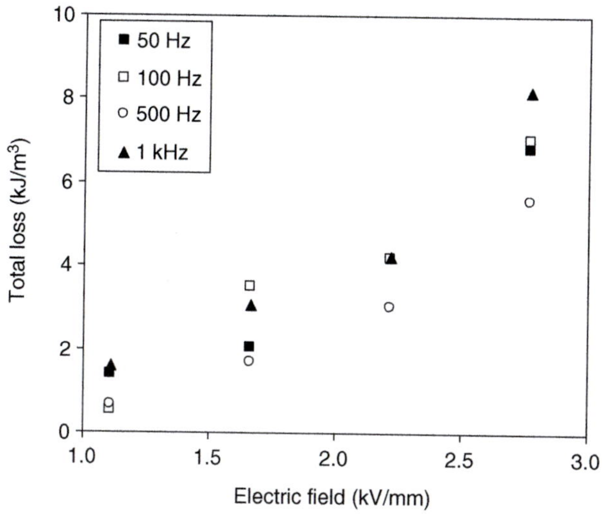

Figure 8.26 Influence of electric field on loss per cycle (u), with drive frequency. *Source:* Pritchard et al. [32] / with permission of Elsevier.

(2.8 kV/mm) is a relatively large value for use in practical applications; therefore, on neglecting those high-field results, u values are broadly between 1 and 3 kJ/m^3. Values for bipolar operation, where full polarisation reversal of the domains occurs, could be higher.

From this case study, it is clear that increasing the driving frequency results in an increase in heating. In addition, the nature of the piezoelectric material will also have an impact on the amount of heating per drive cycle and the selection of ferroelectric materials with more mobile domains will result in a larger amount of loss/hysteresis per drive cycle. Increasing the driving field will also lead to a greater degree of domain motion, leading to a greater amount of loss per cycle and self-heating; see discussions on the polarisation–electric field and strain–electric field loops in Chapter 3. Self-heating is often a concern since heating will facilitate domain motion so that the amount of hysteresis per cycle can gradually increase as the temperature of the actuator rises. Cooling can be employed to improve heat transfer and reduce the level of heating; this can include forced airflow. As an excellent resource on thermal issues related to high-power piezoelectric devices, a Measurement Good Practice Guide is available [40] that aims to provide engineers with an understanding of the challenges, why they occur and how to avoid them.

8.6 Piezoelectric Actuation of Bistable Morphing Structures

There is an increasing interest in the creation of engineering structures that can 'morph', change shape or exhibit large deflections, for example, to provide aerodynamic control. As an example, actuation has been considered to provide changes in shape in micro-air vehicles and drones to provide enhanced levels of manoeuvrability [41, 42] or improve the performance of helicopter rotor blades [43, 44].

We have seen in Chapter 3 that the piezoelectric strain that can be achieved by a piezoelectric ceramic is relatively small (typically up to ~0.1%). In this regard, there has been interest in exploiting bistability, or even multistability, in certain types of engineering structures to amplify the level of strain so that a piezoelectric actuator is simply used to transition an engineering structure between each stable configuration. By exploiting bistability in engineering structures, relatively large displacements and changes in geometry are possible at small levels of power since there is no need to continuously provide a voltage to the piezoelectric actuator.

This case study will examine the behaviour of a bistable carbon fibre–epoxy laminate and its change in shape and deflection due to a piezoelectric strain that is delivered by bonded piezo-fibre-based actuators. For this case study, macro-fibre composite (MFC)-based piezoelectric actuators are employed. These types of piezoelectric actuators are particularly attractive for this application due to their relatively high levels of mechanical flexibility so that they can conform to the range of stable shapes of the structure.

To produce the bistable structure, a non-symmetrical $[0/90]_T$ carbon fibre-reinforced plastic (CFRP) laminate was formed by curing a carbon fibre–epoxy prepreg at high temperature. On cooling to room temperature, relatively large thermal stresses are produced due to the thermal expansion mismatch between the carbon fibre (which has a low thermal expansion coefficient) and the epoxy resin matrix (which has a large thermal expansion coefficient). The thermal stresses produced on forming at an elevated temperature led to the generation of three potential states of equilibrium; see Figure 8.27. The curved geometry

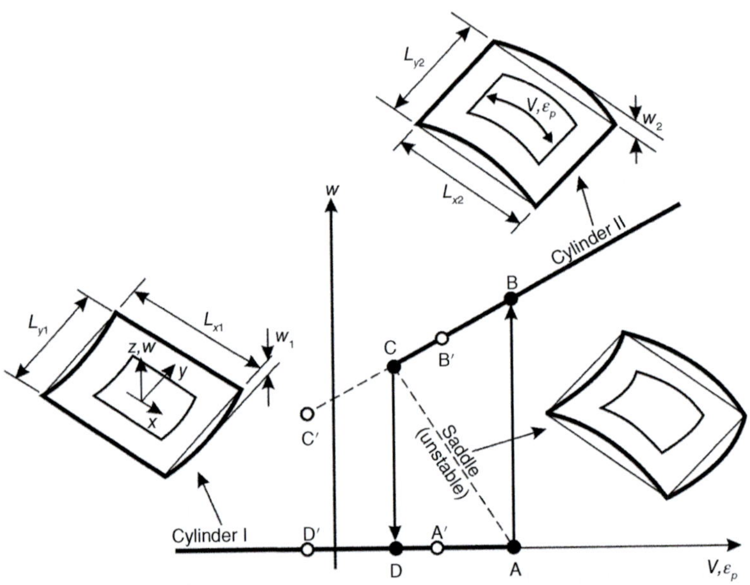

Figure 8.27 Bistable laminate and its two states of equilibrium and schematic of the centre displacement, *w* (*y*-axis), as a function of piezoelectric strain ε_p or potential on the piezoelectric actuator, *V* (*x*-axis). Application of a mechanical load displaces points ABCD to A'B'C'D'. *Source:* Bowen et al. [47] / with permission of Sage publications.

originates from the contrasting thermal expansion coefficients of the individual layers in the CFRP laminate. In reality, the saddle shape in Figure 8.27 is not stable [45–47] so the laminate exhibits one of the two possible cylindrical shapes, resulting in bistable behaviour. An MFC actuator was bonded to the bistable laminate, and a potential, V, was used to induce a piezoelectric strain, ε_p, that offsets the thermally induced strain. The piezoelectric actuator is therefore able to actuate the bistable structure from one state to another and achieve a large change in shape. Figure 8.27 shows how the bistable laminate is able to transition from Cylinder I to Cylinder II via pathway AB and can transition from Cylinder II to Cylinder I by following pathway CD due to the unstable nature of the saddle shape [47].

8.6.1 Composite Structure Manufacture

The unsymmetrical carbon fibre laminate was based on a $[0/90]_T$ lay-up using a CFRP material, which was produced using the conventional approach of laying up the laminate and then autoclaving at elevated temperatures (~125 °C) to cure the epoxy. Figure 8.28 shows an image of the manufactured laminate with a macro-fibre piezoelectric actuator bonded to its surface and the two potential states of the bistable structure. These will be termed 'State I' and 'State II' and are equivalent to Cylinder I and Cylinder II, respectively, in Figure 8.27 [47].

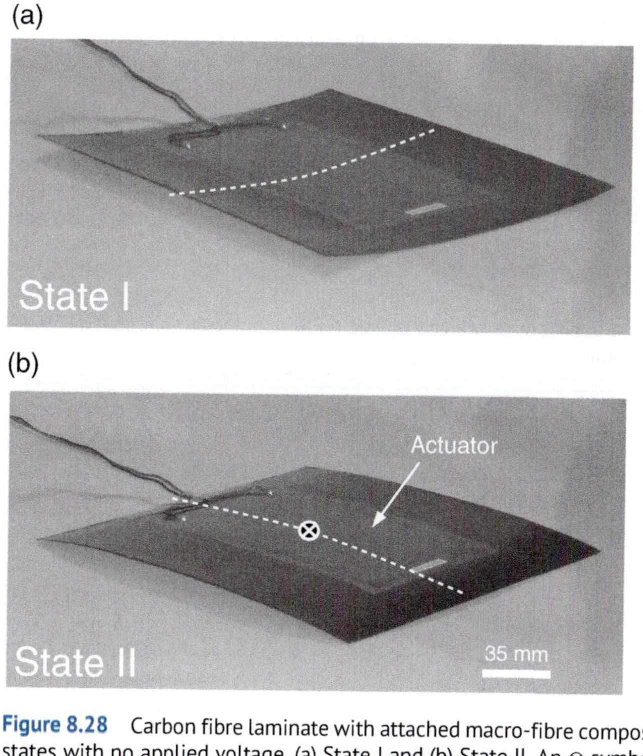

Figure 8.28 Carbon fibre laminate with attached macro-fibre composite (MFC) actuator in two stable states with no applied voltage, (a) State I and (b) State II. An ⊗ symbol shows where any mechanical load was applied. *Source:* Bowen et al. [47] / with permission of Sage Publications.

8.6.2 Actuator Materials and Attachment

The type of actuator used to couple with the bistable carbon fibre laminate is an MFC [47]. This type of piezoelectric actuator is produced from piezoelectric fibres, where an interdigitated electrode (IDE) structure is employed to ensure that the electric field from the applied voltage is aligned along the length of the piezoelectric fibres to exploit the large piezoelectric d_{33} strain coefficient of the PZT-based piezoelectric ceramic used to make the fibres. An MFC with an active piezoelectric area of 85 × 57 mm was used, which covered almost half of the surface area of the bistable composite. The piezoelectric composition selected to produce MFC was a Navy Type II material. As outlined in Chapter 3, this is a soft PZT ceramic that has a high d_{33} piezoelectric strain coefficient to maximise the level of piezoelectric strain. The maximum potential that can be applied to the MFC was 1500 V, due to electrode spacing of 0.5 mm in the IDE pattern; this corresponds to a maximum electric field of 3 kV/mm and maximum free (unloaded) strain of 0.1–0.14%. As the role of the MFC actuator is to achieve snap-through between the two stable states of the laminate, as shown in Figure 8.27, it is worth making a comparison of the magnitude of the piezoelectric strain (ε_p) and the thermally induced strains (ε_t) produced on curing the carbon fibre laminate; it is the thermal strain that produces the two states in Figure 8.28. The thermal expansion coefficient (α) of the CFRP laminate is ~0 °C^{-1} in the carbon fibre orientation but is much higher at ~30 × 10^{-6} °C^{-1} normal to the fibre direction since it is dominated by the larger thermal expansion coefficient of the epoxy matrix in this direction [46]. Since a cure temperature of 125 °C was employed to produce the CFRP laminate, the estimated thermal strain ($\alpha\Delta T$) is ~0.3%; this is larger than the possible piezoelectric strain (~0.1%) [47]. Table 8.4 summarises the properties of the CFRP prepreg and the MFC actuator.

8.6.3 Change in Laminate Shape in Response to Piezoelectric Actuation

The shape of the bistable laminate for applied potentials from 0 to 1500 V is shown in Figure 8.29. Prior to the application of a potential to the MFC, it was possible to position

Table 8.4 Material properties of carbon fibre-reinforced plastic (CFRP) composite and macro-fibre composite (MFC) piezoelectric actuator.

	CFRP HTA (12k) 913	Piezoelectric MFC
E_{11} (GPa)	135	30.3
E_{22} (GPa)	18.5	15.9
G_{12} (GPa)	5.98	5.51
ν_{12}	0.29	0.31
Free strain per volt (10^{-6}/V)	—	0.75–0.90 (low field–high field)
% strain @ 1500 V	—	0.11–0.135

Source: Bowen et al. [47] / with permission of Sage publications.

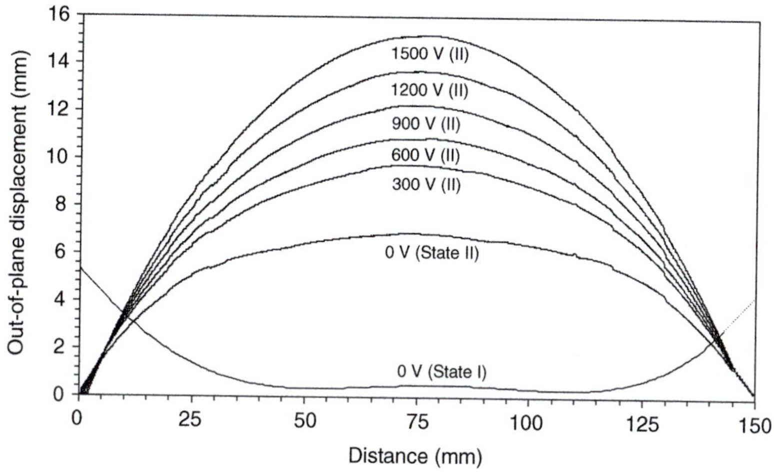

Figure 8.29 Displacement profile at a carbon fibre laminate for a range of voltage levels. Transitions between State I and State II were irreversible. *Source:* Bowen et al. [47] / with permission of Sage publications.

the bistable laminate in State I or State II, where the side profiles are identified as '0 V (State I)' and '0 V (State II)' in Figure 8.29. If the bistable structure was in '0 V (State I)', it was possible to induce a transition from stable State I to State II at potentials of 100–150 V. When the potential was then gradually increased to 1500 V, the degree of laminate displacement in State II increased; see Figure 8.29. When the potential was removed and returned to 0 V, the degree of laminate curvature was then reduced, but the laminate stayed in stable State II. The measured displacements were relatively large, and deflections up to 15 mm were achieved; these large deflections are clearly much larger than the free displacement of an MFC. For example, if the length of the MFC was 85 mm and the maximum strain at 1500 V is ~0.1–0.14%, this is ~85–120 μm. However, the change in shape was irreversible, and the laminate could not be returned to its original state (State I), so reversible transitions between the stable states could not be achieved.

One approach to attaining reversible snap-through is to couple the piezoelectric strain with some forms of external loads that provide a mechanical bias to the structure. Figure 8.30 shows an image of the bistable composite with a simple mass placed at its centre;

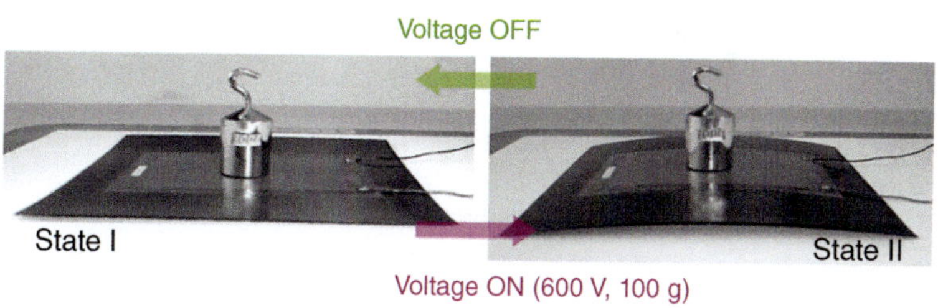

Figure 8.30 Changes between State I and State II using a load (100-g mass). Application of 600 V to the piezoelectric led to snap-through from State I to State II. *Source:* Bowen et al. [47] / with permission of Sage Publications.

the mass was large enough to transition the bistable structure from State II to the more flat and stable State I when 0 V was applied to the piezoelectric; see Figure 8.30, left image [47]. When a potential of 600 V was applied to the piezoelectric actuator, snap-through was possible from State I to State II, as shown in Figure 8.30, right image. When the potential applied to the piezo-actuator was removed again, the laminate transitioned from State II to State I, back to Figure 8.30, left image. As a result, the combination of a mechanical load and a piezoelectric strain provided a route to achieve reversible snap-through [47].

The changes in geometry/profile, degree of deflection and the drive potential required to ensure that transitions of the bistable laminate from one stable shape to another depend on the level of mechanical bias. Measurements were therefore undertaken for a range of bias masses (51–408 g) that were applied to the centre of the laminate. Figure 8.31 summarises the displacement of the centre of the bistable laminate as the potential to the actuator increases to 1500 V and subsequently decreases back to 0 V. When a small mass was employed (51 g, 0.5 N), a transition from State I to State II occurred at 200 V, leading to a large deflection at potentials between 100 and 200 V. When the potential applied to the piezoelectric actuator was increased up to 1500 V, an almost linear increase in deflection with drive potential was observed. On the returning 'downward' voltage cycle to 0 V, there is a continual decrease in displacement with a decrease in applied potential. Figure 8.31 shows that some hysteresis is observable in the displacement–voltage curves between the 'upward' (0–1500 V) and 'downward' (1500–0 V) cycles, and this may be due to ferroelectric hysteresis and creep of the piezoelectric actuators due to domain motion in the relatively soft ferroelectric material [47–49]; both hysteresis and creep have been discussed in Chapter 3. As a

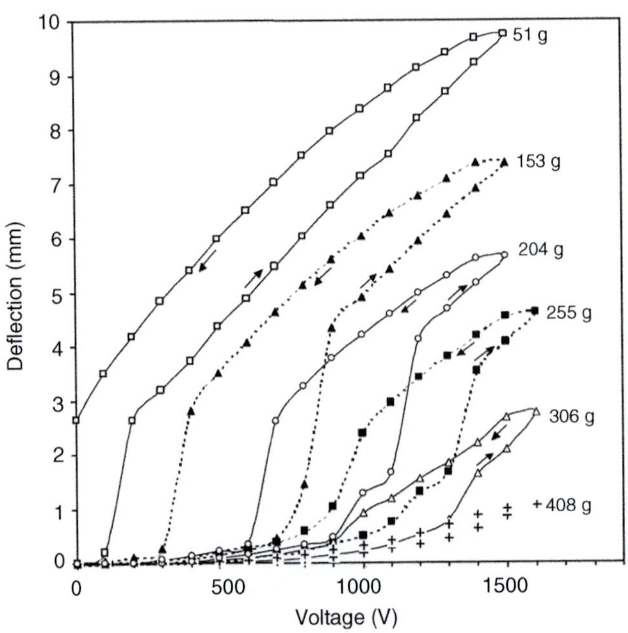

Figure 8.31 Laminate centre deflection with the applied voltage increasing and then reducing in 100-V steps, with applied loads of 0.5 N (51 g)–4 N (408 g). *Source:* Bowen et al. [47] / Sage publications.

low mass of 51 g was not able to result in transitions from State II to State I during the down cycle, a high deflection is maintained when the applied potential is removed; this is shown in Figure 8.29.

This case study has shown that it is possible to exploit bistability to produce large displacement in materials and structures. The piezoelectric strain is used to offset thermal strains in an engineering structure to transition the bistable structure between two stable states. Tuning of the transitions between the stable states is possible by tailoring the thermal strains, piezoelectric strain and application of mechanical load to also provide an additional mechanical strain.

8.7 Force Sensors, Shear Sensors and Hydrophones

In this final case study, we examine the use of freeze casting to produce a range of piezocomposite materials with different properties and relevant figures of merit. We have seen in Chapters 2 and 5 that there has been an attempt to reduce the permittivity of piezoelectric materials by introducing low-permittivity pores, or a low-permittivity polymer phase, into a dense ferroelectric ceramic to create a piezocomposite material. In this regard, there is a need to strike a balance between the required mechanical properties, such as strength and stiffness, maintaining a sufficient level of piezoelectric activity and the level of porosity or polymer that is introduced into the material structure. This is of particular importance for porous ceramics that contain pores which are distributed randomly through the microstructure of the material; a random distribution of pores is often achieved when using an organic pore-forming inclusion, which volatilises during a high-temperature heat treatment. We show in this case study that porosity can lead to an increase in the d_h hydrostatic charge coefficient, primarily due to a reduction in d_{31} coefficient, while maintaining a high d_{33} coefficient. An increase in g_h is also observed, which is a result of an increase in the d_h and a decrease in permittivity, indicating potential for high-sensitivity hydrophones that operate under low-frequency hydrostatic conditions. For sensor applications that operate under uniaxial loads, such as accelerometers and pressure sensors in Figures 5.1, 5.6 and 5.7, the ability to achieve a high d_{33} or d_{31} piezoelectric charge coefficient, along with a low permittivity, leads to materials with large g_{ij} voltage coefficients, since $g_{33} = d_{33}/\varepsilon_{33}^T\varepsilon_0$ and $g_{31} = d_{31}/\varepsilon_{33}^T\varepsilon_0$, so that they are able to generate a large electric field per unit stress for sensing.

8.7.1 Freeze Casting to Produce Porous Ceramics

Freeze casting initially involves the dispersion of a ceramic powder, such as PZT particles, in water, which is then ball-milled to generate a homogeneous and fine ceramic–water suspension. Freeze casting involves directional freezing of the PZT ceramic particles, which are dispersed in the water, as shown in Figure 8.32a; the process initially gained attention in the early 2000s for processing bioceramics. As the ceramic–water suspension is directionally solidified from its base, ice dendrites form and the ceramic particles are pushed into the space between the ice to create an aligned ceramic structure; see Figure 8.32b. Once frozen, the ice–ceramic mix is freeze-dried and then sintered at high temperature to create a porous

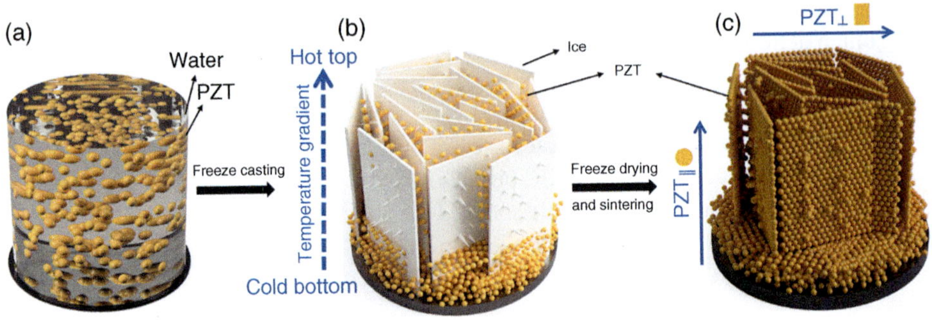

Figure 8.32 Diagram of the freeze casting process. (a) Water-based ceramic suspension; PZT is lead zirconate titanate, (b) directional (upward) freezing of the water–ceramic suspension from a cold base and (c) removal of ice crystals using freeze drying, leaving behind a porous ceramic microstructure that is then sintered. The direction of freezing (bottom to top) and poling direction for materials that are poled parallel to the freezing direction ($PZT_{\parallel}$) and at right angles to the freezing direction ($PZT\perp$). *Source:* Zhang et al. [50] / with permission of Elsevier.

ceramic, as shown in Figure 8.32c. The advantage of this process is that the pores are highly aligned, which can be beneficial for achieving a high level of polarisation, and therefore lead to large piezoelectric d_{ij} coefficients.

In this case study, a porous PZT material was produced via freeze casting, where the ceramic was poled in two different directions, namely parallel and perpendicular to the freezing direction. These two poling directions are described by the terms $PZT_{\parallel}$ (parallel poling, relative to freezing direction) and $PZT\perp$ (perpendicular poling, relative to freezing direction). This ability to change the poling direction after manufacturing the materials demonstrates the design flexibility of a ferroelectric material since the polarisation direction can be tailored according to the poling direction [50].

Figure 8.33 shows scanning electron microscopy images of the resulting PZT ceramic microstructure parallel to the freezing direction. The pore fraction is tailored by changing the amount of ceramic powder added to the water–ceramic suspension that is frozen; here, the porosity volume fraction has been varied from 20 to 60 vol.%. For the 20 vol.% ceramic (i.e. freezing a suspension with a low ceramic loading level), a number of side dendrites are observed which are produced during the direction freezing process in Figure 8.32. The dendrites can be seen to connect to adjacent ceramic layers that aid in improving the mechanical properties since they connect the individual lamella together.

Figure 8.34 summarises the properties of the porous freeze-cast ceramic materials [50]. By changing the ceramic loading level in the water–ceramic suspension for pore volume fractions ranging between 20 and 60 vol.%, Figure 8.34a demonstrates that $PZT\perp$ has a reduced $-d_{33}/d_{31}$ and therefore a lower level of anisotropy in the d_{ij} coefficient compared to the dense PZT material for the range of pore fractions examined. The porous $PZT_{\parallel}$ has a larger anisotropy factor ($-d_{33}/d_{31}$) than the dense PZT and $PZT\perp$. A large anisotropy factor can be of benefit to hydrostatic coefficients since it means the d_{33} coefficient is larger than the d_{31} coefficient, thereby indicating a decoupling of the two piezoelectric coefficients that can lead to an improved d_h coefficient (see Chapter 5); this is shown in Figure 8.34b. When the material is poled perpendicular to the freezing direction, namely $PZT\perp$, the material

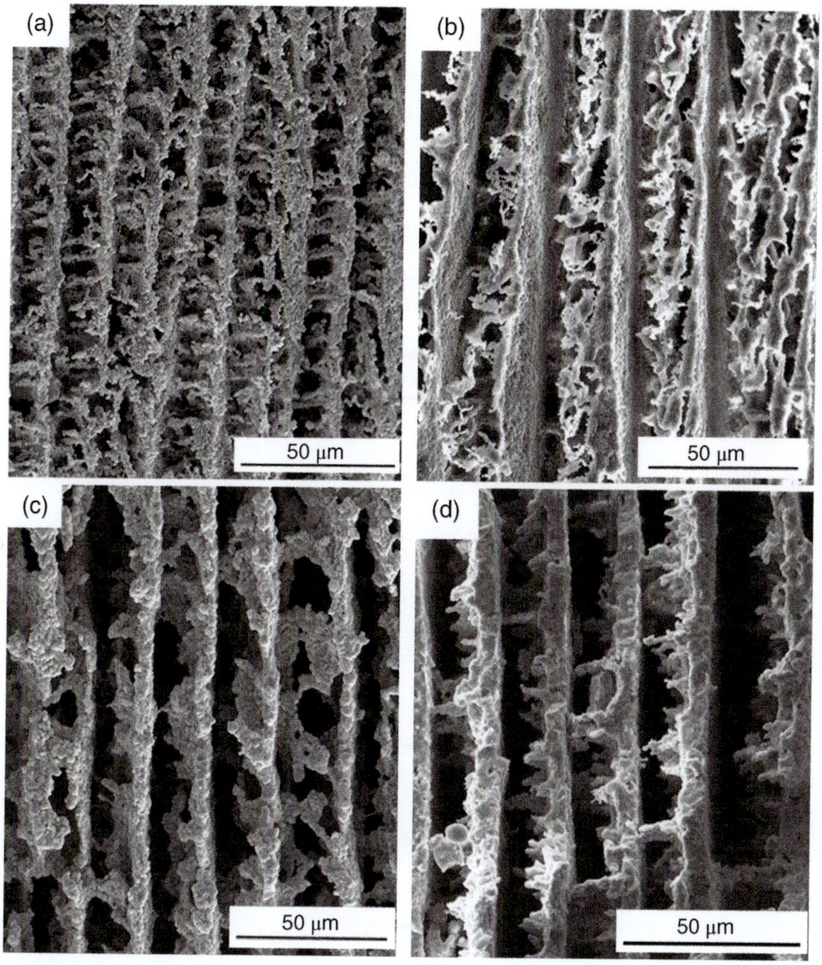

Figure 8.33 Scanning electron microscopy micrographs of porous PZT ceramic formed by freezing casting; freezing direction is upwards; pore volume fractions of (a) 20, (b) 30, (c) 40 and (d) 50 vol.%. *Source:* Zhang et al. [50] / with permission of Elsevier.

exhibits a gradual reduction in d_h as the pore fraction is increased, due to a low d_{33} of these materials. In contrast, when the material is poled along the freezing direction, $PZT_{\parallel}$, the d_h gradually increases with increasing porosity.

While the anisotropy factor is increasing with pore fraction for $PZT_{\parallel}$, the increase in d_h compared to the dense material is relatively modest due to the decrease in d_{33} with pore volume fraction. The introduction of porosity clearly leads to a reduction in the permittivity of both the $PZT_{\parallel}$ material (Figure 8.34c), and the $PZT\perp$ material (Figure 8.34d). A greater reduction in permittivity is observed for $PZT\perp$ due to the reduced connectivity of the ceramic in the direction perpendicular to the freezing direction [50].

The freeze-cast *PZT* materials exhibited good g_h values that are up to 8.5 times greater than the dense ceramic material; see Figure 8.34b. For example, the hydrostatic voltage

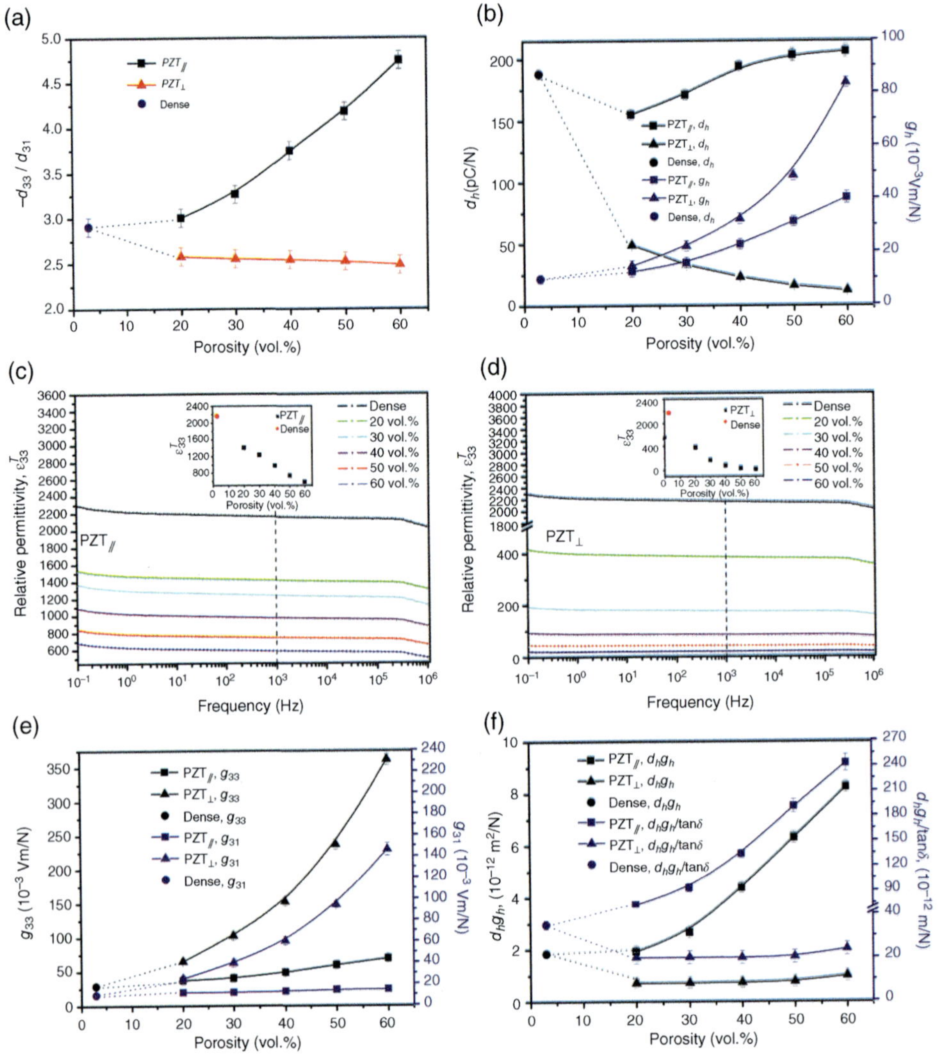

Figure 8.34 (a) $-d_{33}/d_{31}$ anisotropy factor with pore fraction, (b) hydrostatic charge (d_h) and voltage coefficients (g_h) with pore fraction, (c, d) relative permittivity (ε_{33}^T), (e) g_{33} voltage coefficient with pore fraction and (f) hydrostatic figure of merits (d_hg_h and d_hg_h/tanδ) of the freeze-cast PZT for a range of pore volume fractions. Data for dense PZT material are provided for comparison [50].

coefficient $g_h = 40\times 10^{-3}$ Vm/N for $PZT_{\|}$ (poled parallel to the freezing direction) with $g_h = 84\times 10^{-3}$ Vm/N for $PZT\perp$ (poled perpendicular to the freezing direction) for a porosity volume fraction of 60%. The dense PZT performance is significantly lower, where $g_h = 9 \times 10^{-3}$ Vm/N. The large values of the hydrostatic coefficient are a result of the lower permittivity of the porous piezoelectric ceramics, which are shown in Figures 8.34c and d, where the permittivity is relatively constant with frequency but decreases in magnitude rapidly with increasing pore fraction. These results indicate the potential for these materials as high-sensitivity hydrophones.

The introduction of porosity and a corresponding reduction in permittivity also lead to the large g_{33} and g_{31} piezoelectric voltage coefficients in Figure 8.34e. The g_{33} and g_{31} voltage coefficients of $PZT\perp$ were up to 5 and 10 times larger than $PZT_{||}$ and up to 12 and 15 times larger than the dense PZT ceramic, respectively. These large voltage coefficients indicate that $PZT\perp$ may be of interest as a force, strain, acceleration or pressure sensor [50]; the attractive voltage coefficients result from the very low permittivity of the material in this direction.

Figure 8.34d indicates that the calculated hydrophone figures of merit, $d_h{\cdot}g_h$ and $d_h{\cdot}g_h/\tan\delta$ for $PZT_{||}$ increase with the pore volume fraction; the calculated figures of merit are significantly greater than the dense PZT ceramic. These attractive hydrostatic properties are achieved since the material has (i) a high level of piezoelectric activity (leading to the relatively large piezoelectric d_{ij} coefficients), (ii) a high anisotropy factor ($-d_{33}/d_{31}$) and, in particular, (iii) a low relative permittivity due to the presence of low-permittivity pores in the microstructure. The lower density and stiffness of the porous materials may also provide benefits in terms of better acoustic impedance matching with water; see Chapter 5.

Table 8.5 compares the hydrostatic properties of the freeze-cast materials with composites produced by other routes. The materials produced by directional freeze casting provide attractive d_h, g_h and $d_h{\cdot}g_h$ values compared to other routes to create composites. The freeze casting process is also relatively simple to implement and can be scaled up; it is also of interest to compare with the hydrostatic properties of dense PZT ceramic in Table 8.5.

Table 8.5 Summary of materials properties and hydrostatic coefficient of PZT composite materials formed by a range of processing methods. BURPS indicates the Burn-out of polymer spheres method.

Production method		Composite	Connectivity	PZT vol. %	d_h (pC/N)	g_h (10^{-3} Vm/N)	$d_h{\cdot}g_h$ (10^{-15} m^2/N)
Freeze cast	$PZT\|\|$	PZT–air	2–2	40	206	40.1	8260
	$PZT\perp$				12	83.5	1000
BURPS		PZT–air	3–0	54.5	~35	~16.4	~570
			3–1	59	—	~42	~5000
			3–3	~60	102	72	7300
			3–0	68	—	~48	5000
Ionotropic gelation		PZT–air	3–1	~44	~222	—	~5600
Pore-forming agent		PZT–epoxy	3–1	45–60	60	69	~4010
Polymer injection (PZT rods in polymer)		PZT–polymer	1–3	40	~95	~18	~1710
Direct-write		PZT–polymer	2–2	~30	<190	<0.38	<720
Dice-and-fill		PZT–polymer	1–3	40	~135	~37	~5000
		PZT–cement	2–2	40	<100	<0.2	<20

Source: Lee et al. [21] / Sage publications.

8.7.2 Fabrication of Strain Sensor (d_{31}-Mode)

Piezoelectric ceramic formed by freeze casting was utilised as a d_{31}-mode sensor and attached to an automotive tyre, as indicated in Figure 8.35. A portion of a car tyre was tested using an Instron 1342 fatigue tester, and a strain gauge was also positioned on the central axis of the tyre to measure the strain and compare it with the voltage output of the piezoelectric sensor (Figure 8.35a). The tyre was deformed using a compressive load that was cycled at a range of frequencies, with a maximum frequency of 10 Hz, which can be typical

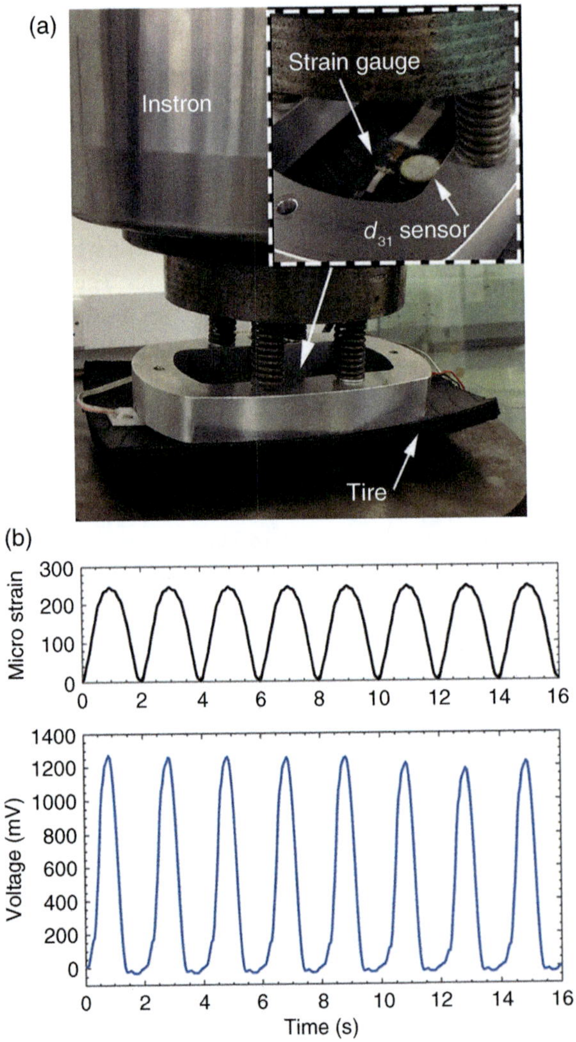

Figure 8.35 d_{31}-mode piezoelectric sensor located on an automotive tyre. (a) An experimental configuration for dynamic testing, including the use of a strain gauge as a reference. *Source:* [51] / with permission of Royal Society of Chemistry. (b) Strain at sidewall of the tyre during deformation (measured from strain gauge) and corresponding open-circuit voltage of the d_{31}-mode piezoelectric sensor. *Source:* Adapted from Xie et al. [51].

of a tyre. Figure 8.35b provides the strain gauge measurements while applying a cyclic compression of 0 to 3.5 kN, along with the measured open-circuit voltage of the d_{31}-mode piezoelectric sensor that was positioned at a similar position. The measured piezoelectric response, determined by the measured piezoelectric voltage, is seen to be in good agreement with the strain gauge data, where the maximum open-circuit voltage of ~1200 mV corresponds to a strain of ~250 μ strain [51].

8.7.3 d_{33}-Mode and d_{15}-Mode Piezocomposite Sensors

A uniaxial compressive/tensile strain sensor that operated in d_{33}-mode and a shear sensor that operated in d_{15}-mode were manufactured using porous piezoelectric ceramic; the sensors were able to detect both mechanical taps and shearing. A dedicated, custom geometry, freezing mould was produced by 3D printing to give the required device structure, which included a number of ceramic sensing pillars, as shown in Figure 8.36a. A diagram of the

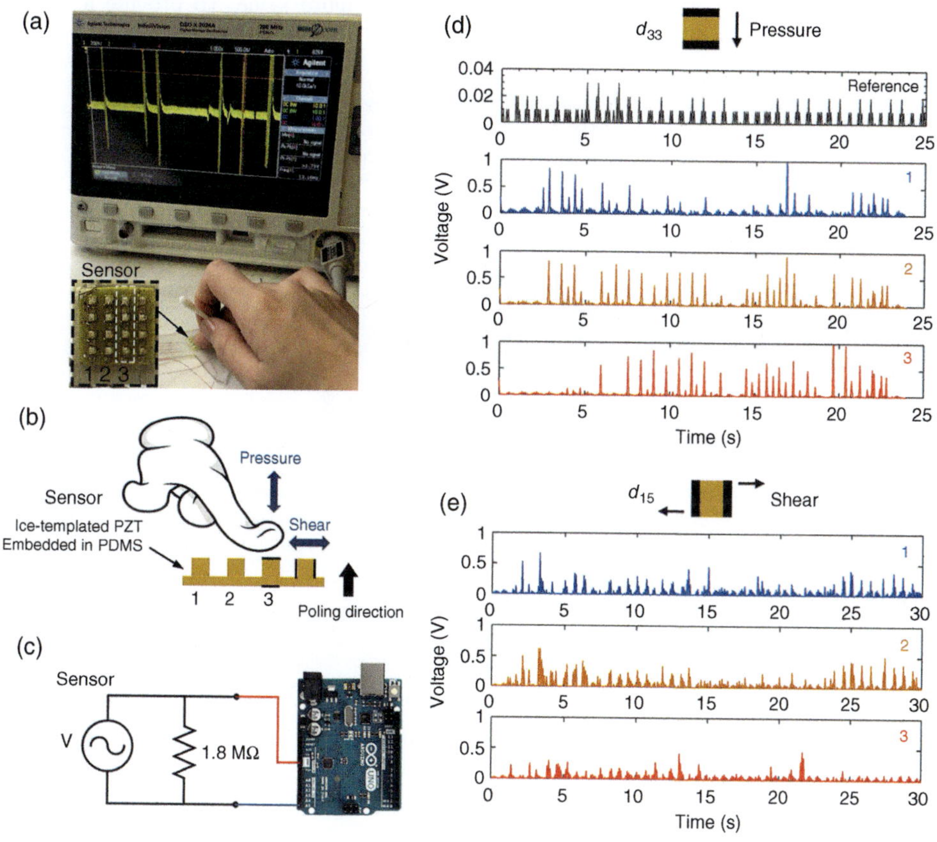

Figure 8.36 (a) Image of d_{33}-mode and d_{15}-mode piezo-sensing array. *Source:* [51] / with permission of Royal Society of Chemistry. (b) diagram of tapping measurement, (c) diagram of measurement circuit, (d) d_{33}-mode response on loading by mechanical tapping (shown with a reference signal without touching, response of sensor array with light finger-tapping and shearing) and (e) response of d_{15}-mode piezoelectric shear sensor. *Source:* Adapted from Xie et al. [51].

experimental method to examine the sensor response to an applied pressure and/or shear is outlined in Figure 8.36b. The individual piezoelectric pillars had a common bottom electrode, whereas the d_{33}-mode uniaxial pressure sensor used four individual electrodes located on top of the piezoelectric pillars. Mechanical tapping on a line of pillars was used to operate/test the sensor in d_{33}-mode, and the generated piezoelectric voltage was measured using a high-input-impedance (>100 TΩ) electrometer [51]. A high input impedance ensures an accurate measurement of the open-circuit voltage since the use of a low-impedance oscilloscope (e.g. one with an input impedance of only 1 MΩ) can lead to an inaccuracy in the open-circuit voltage measurement due to discharge of the piezoelectric charge through the measurement equipment; in such a case, the open-circuit voltage would be lower than expected.

To characterise the response of the d_{33}-mode and d_{15}-mode sensor arrays to either mechanical tapping by a downward force or shearing, the sensor was connected to an Arduino board. Figure 8.36c is a diagram of the measurement circuit, and the sensor voltage–time response in d_{33}-mode (downward pressure) is indicated in Figure 8.36d. When in d_{15}-mode (shearing), voltage–time data are shown in Figure 8.36e. To provide a 'zero'

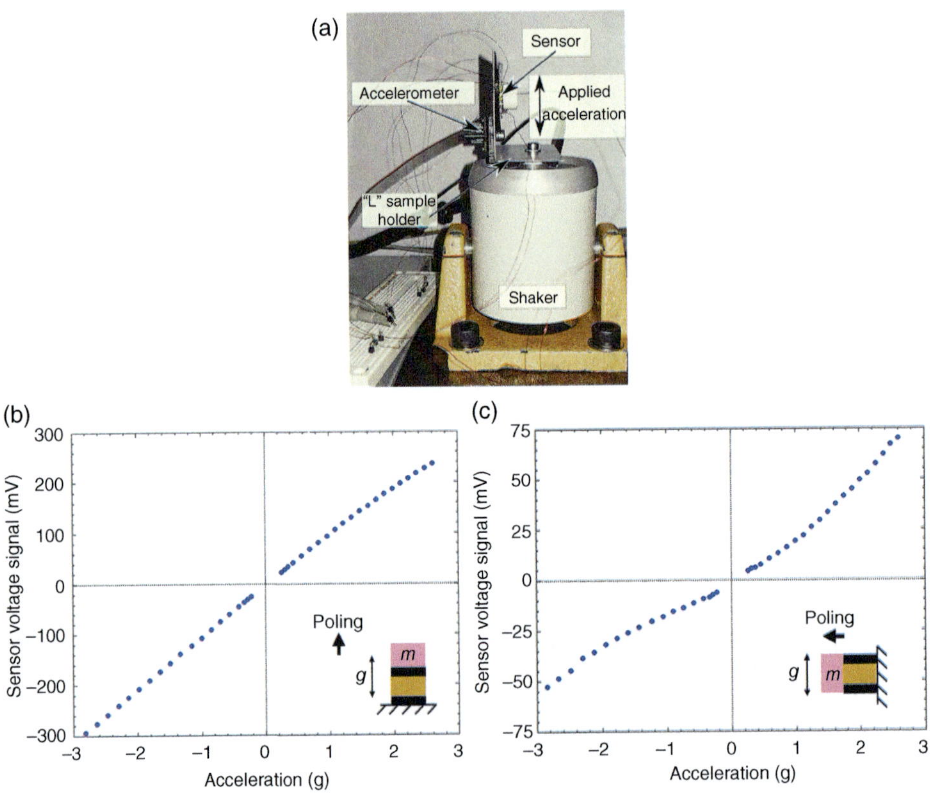

Figure 8.37 Sensitivity of piezocomposite sensors as a function of acceleration. (a) Configuration to determine sensor response. (b, c) Voltage signal for a d_{33}-mode piezoelectric sensor and a d_{15}-mode sensor versus acceleration, respectively (m is a seismic mass attached to the sensor; the poling direction and the direction of acceleration provided by the shaker are indicated in the inset of the figure). *Source:* Adapted from Xie et al. [51] / Royal Society of Chemistry.

reference, the response of the sensor with no mechanical tapping or shearing is shown in Figure 8.36d. A small level of noise, less than 20 mV, is observed and with additional amplification of the piezoelectric signal or filtering, as outlined in Chapter 7, the noise level can be reduced further [51].

The sensitivity of the d_{33}-mode pressure and d_{15}-mode shear sensors was examined in response to a range of accelerations that were provided by attaching the sensors to a shaker system, as shown in Figure 8.37a. To produce a simple d_{33}-mode accelerometer, an end mass was bonded to the top of the porous piezocomposite, similar to the sensor structures outlined in Chapter 5. Likewise, a d_{15}-mode sensor was also formed by bonding an end mass. The locations of the seismic masses for the two types of piezoelectric sensors are shown in Figure 8.37b. A reference accelerometer was also used as a control, and the shaker system was used to provide accelerations that ranged between −3 and +3 g (~ +/− 29.5 m/s^2).

The piezo-sensor performance is outlined in Figures 8.37b and c for the d_{33}-mode sensor and d_{15}-mode sensor. The d_{33}-mode and d_{15}-mode sensors are observed to provide a good level of linearity with acceleration, although the d_{33}-mode sensor provides the most linear response over the range of accelerations tested. The level of mechanical loading for the d_{15}-mode sensor is larger, which can lead to non-linear behaviour (see Chapter 3). For an acceleration range of 0.5–3 g, the d_{15}-mode sensor provides a sensitivity of ~27 mV/g (~440 mV/N), with the d_{33}-mode sensor having a higher sensitivity of ~90 mV/g (~690 mV/N) [51].

8.8 Concluding Remarks

The chapter has provided an overview of example applications for piezoelectric materials, with a focus on piezoelectric actuation and piezocomposite sensing. The examples aim to demonstrate that in order to ensure appropriate material selection and to use the materials correctly, an understanding of the underlying theory of these fascinating materials, as outlined in Chapters 2–7, is essential.

References

1 Schöppe, D., Zülch, S., Hardy, M. et al. (2008). Delphi common rail system with direct acting injector. *MTZ Worldwide* 69 (10): 32–38.

2 Persson, L.J. 2016. Design and control of piezoelectric actuation for hydraulic valves. PhD thesis. University of Bath.

3 Plummer, A. 2016. Electrohydraulic servovalves – past, present, and future. *In: 10th International Fluid Power Conference, IFK2016*. Dresden.

4 Boyar. R. E., Johnson, B. A. and Schmid, L. 1955. Hydraulic Servo Control Valves Part 1. *WADC Technical Report 55-29*, Wright-Patterson Air Force Base, Ohio.

5 Johnson, R., Stahl, R. and Walters, G. (Textron Inc) Non-magnetic electro-hydraulic transfer valve. Patent US2928409, filed January 1955 and issued on March 1960.

6 Murrenhoff, H. 2003. Trends in valve development. O + P (Ölhydraulik und Pneumatik) 46, Nr. 4.

7 Jeon, J., Nguyen, Q.-H., Han, Y.-M., and Choi, S.-B. (2013). Design and evaluation of a direct drive valve actuated by piezostack actuator. *Advances in Mechanical Engineering* 5.

8 Karunanidhi, S. and Singaperumal, M. (2010). Mathematical modelling and experimental characterization of a high dynamic servo valve integrated with piezoelectric actuator. *Proceedings of the Institution Mechanical Engineers, PartI: Journal of Systems and Control Engineering* 224: 419–435.

9 Sangiah, D., Plummer, A.R., Bowen, C., and Guerrier, P. (2014). A novel piezohydraulic aerospace servovalve. Part I : Design and modelling. *Proceedings of the Institution of Mechanical Engineers, Part I: Journal of Systems and Control Engineering* 227 (4): 371–389.

10 Persson, L.J., Plummer, A.R., Bowen, C. and Brooks, I. 2015. Design and modelling of a novel servovalve actuated by a piezoelectric ring bender. *Proceedings of the ASME/Bath Symposium on Fluid Power and Motion Control*, Chicago, (October 2015).

11 Persson, L.J., Plummer, A.R., Bowen, C., and Brooks, I. (2016). A lightweight, low leakage piezoelectric servovalve. In: *Proceedings of International Conference on Recent Advances in Aerospace Actuation Components and Systems 2016 (R3ASC'16)*. Toulouse.

12 Reichert, M. and Murrenhoff, H. 2006. New concepts and design of high response hydraulic valves using piezo-technology. *Power Transmission and Motion Control Symposium*, Bath, (September 2006).

13 Cui, Q., Liu, C., and Zha, X.F. (2008). Simulation and optimization of a piezoelectric micropump for medical applications. *The International Journal of Advanced Manufacturing Technology* 36: 516.

14 R. Rammer and P. Jänker 2007. Potential of piezo hydraulic actuators for aerospace. Presented at the Recent Advances in Aerospace Actuation Systems and Components, Toulouse, France.

15 Chaudhuri, J.-H.Y. and Wereley, N.M. (2009). Design, test and model of a hybrid magnetostrictive hydraulic actuator. *Smart Materials and Structures* 18: 1–21.

16 Chaudhuri and Wereley, N.M. (2011). Compact hybrid electrohydraulic actuators using smart materials: a review. *Journal of Intelligent Material Systems and Structures* 22: 1–37.

17 John, S., Cadou, C., Yoo, J.H., and Wereley, N.M. (2006). Application of CFD in the design and analysis of a piezoelectric hydraulic pump. *Journal of Intelligent Material Systems and Structures* 17: 967–979.

18 John, S., Sirohi, J., Wang, G., and Wereley, N.M. (2007). Comparison of piezoelectric, magnetostrictive, and electrostrictive hybrid hydraulic actuators. *Journal of Intelligent Material Systems and Structures* 18: 1035–1048.

19 Nosse, D.T. and Dapino, M.J. (2007). Magnetorheological valve for hybrid electrohydrostatic actuation. *Journal of Intelligent Material Systems and Structures* 18: 1121–1136.

20 Denkena and Simon, S. (2008). Computer-aided analysis of flow-characteristics of a micro-positioning system. *Production Engineering Research and Development* 2.

21 Lee, G., Or, S.W., and Carman, G.P. (2004). Design of a piezoelectric-hydraulic pump with active valves. *Journal of Intelligent Material Systems and Structures* 15: 107–115.

22 Sirohi, J. and Chopra, I. (2003). Design and development of a high pumping frequency piezoelectric-hydraulic hybrid actuator. *Journal of Intelligent Material Systems and Structures* 14: 135–147.

23 Henderson J, Plummer A, Johnston N and Bowen C 2013. The influence of passive valve characteristics on the performance of a piezo pump. *ASME/BATH 2013 Symposium on Fluid Power and Motion Control*, Sarasota, (6–9 October 2013).

24 Plummer, A, Manning, W and Levesley, M 2000. Modelling and control of vibration in a smart structure. *Workshop on Power Transmission and Motion Control (PTMC)*, University of Bath, (2000).

25 Uchino, K. (1993). Ceramic actuators: principles and applications. *Materials Research Society Bulletin* 18: 42–47.

26 K. Uchino and S. Takahashi, in M.W. Roberos and J. White (eds.), Multilayer ceramic actuators, *Current Opinion in Solid State and Materials Science*, 1 (1996) 698–705.

27 C. Schuh, Th. Steinkopff, A. Wolff and K. Lubitz, Piezoceramic multilayer actuators for fuel injection systems in automotive area, *Smart Structures and Materials 2000: Active Materials, Proceedings of SPIE,* 3992 (2000) pp. 165–175.

28 Wheat, T.A., Ahmad, A., and Minhas, R.K. (1999). Ceramic actuators: materials. *Journal of the Canadian Ceramic Society* 69: 56–61.

29 de Vries, J.W.C. (1999). Functional behaviour of multilayer actuators with inactive parts. *Sensors and Actuators A* 72: 251–255.

30 Ochi, A., Takahashi, S., and Tagami, S. (1985). Temperature characteristics for multilayer piezoelectric ceramic actuator. *Journal of Applied Physics* 24 (S3): 209–212.

31 Härdtl, K.H. (1982). Electrical and mechanical losses in ferroelectric ceramics. *Ceramics International* 8: 121–128.

32 Pritchard, J., Ramesh, R., and Bowen, C.R. (2004). Time-temperature profiles of multi-layer actuators. *Sensors and Actuators A: Physical* 115: 140–145.

33 Zheng, J., Takahashi, S., Yoshikawa, S. et al. (1996). Heat generation in multilayer piezoelectric actuators. *Journal of the American Ceramic Society* 79: 3193.

34 Nagata, K. and Kinoshita, S. (1995). Relationship between lifetime and multilayer ceramic actuator temperature. *Japanese Journal of Applied Physics* 32: 5266–5269.

35 Thomgrueng, J., Tsuchiya, T., and Nagata, K. (1998). Lifetime and degradation mechanisms of multilayer ceramic actuator. *Japanese Journal of Applied Physics* 37: 5306–5310.

36 Lente, M.H. and Eiras, J.A.L. (2000). Interrelationship between self-heating and ferroelectric properties in PZT ceramics during polarization reorientation. *Journal of Physics. Condensed Matter* 12: 5939–5950.

37 Kunkel, H.A., Locke, S., and Pikeroen, B. (1990). Finite-element analysis of vibrational modes in piezoelectric ceramic disks. *IEEE Transactions on Ultrasonics, Ferroelectrics, and Frequency Control* 37: 316–328.

38 N. Abboud, J. Mould, G. Wojcik 1997. Thermal generation, diffusion and dissipation in 1-3-piezocomposite sonar transducers: Finite element analysis and experimental measurements, *1997 IEEE Ultrasonics Symposium Proceedings,* VOLS 1 & 2, Editors: S.C.Schneider, M.Levy, B.R.McAvoy, (1997), pp. 895–900.

39 Uchino, K. and Hirose, S. (2001). Loss mechanisms in piezoelectrics: how to measure different losses separately. *IEEE Transactions on Ultrasonics, Ferroelectrics, and Frequency Control* 48: 307 321.

40 Mark Stewart and Markys G Cain 2005. Measurement and modelling of self-heating in piezoelectric materials and devices. *A National Measurement Good Practise Guide, No. 84.* Engineering and Process Control Division, National Physical Laboratory, ISSN 1368–6550 (2005)

41 Abdulrahim, M., Garcia, H. and Lind, R. 2004. Flight testing a micro air vehicle using morphing for aeroservoelastic control, *45th AIAA/ASME/ASCE/AHS Structures, Structural Dynamics, and Materials Conference*, Palm Springs, California, USA, AIAA Paper 2004-1674.

42 Abdulrahim, M., Garcia, H., and Lind, R. (2005). Flight characteristics of shaping the membrane wing of a micro air vehicle. *Journal of Aircraft* 42 (1): 131–137.

43 Cesnik, C.E.S., Mok, J., Parikh, A.S. and Shin, S. 2004. Optimization design framework for integrally twisted helicopter blades *45th AIAA/ASME/ASCE/AHS/ASC Structures, Structural Dynamics, and Materials Conference*, Palm Springs, California, USA, AIAA Paper 2004-1761.

44 Shin, S., Cesnik, C.E.S., and Hall, S.R. (2005). Closed-loop control test of the NASA/Army/MIT active twist rotor for vibration reduction. *Journal of the American Helicopter Society* 50 (2): 178–194.

45 Hufenbach, W., Gude, W., and Kroll, L. (2002). Design of multistable composites for application in adaptive structures. *Composites Science and Technology* 62 (16): 201–2207.

46 Potter, K.D. and Weaver, P.M. (2004). A concept for the generation of out-of-plane distortion from tailored FRP laminates. *Composites Part A: Applied Science and Manufacturing* 35: 1353–1361.

47 Bowen, C.R., Butler, R., Jervis, R. et al. (2007). Morphing and shape control using unsymmetrical composites. *Journal of Intelligent Material Systems and Structures* 18 (1): 89–98.

48 Fett, T. and Thun, G. (1998). Determination of room-temperature tensile creep of PZT. *Journal of Materials Science Letters* 27: 1929–1931.

49 Jung, H., Shim, J.Y., and Gweon, D. (2001). Tracking control of piezoelectric actuators. *Nanotechnology* 12: 14–20.

50 Zhang, Y., Roscow, J., Xie, M., and Bowen, C. (2018). High piezoelectric sensitivity and hydrostatic figures of merit in unidirectional porous ferroelectric ceramics fabricated by freeze casting. *Journal of the European Ceramic Society* 38: 4203–4211.

51 Xie, M., Zhang, Y., Kraśny, M.J. et al. (2018). Flexible and active self-powered pressure, shear sensors based on freeze casting ceramic–polymer composites. *Energy & Environmental Science* 10: 2919–2927.

Index

Piezoelectric Materials: Properties, Applications and Devices, First Edition.
Christopher R. Bowen and Andrew R. Plummer.

Printed and bound by CPI Group (UK) Ltd, Croydon, CR0 4YY
17/09/2026
14964297-0001